Climate Change Policy Failures

Why Conventional Mitigation Approaches Cannot Succeed

Climate Change Policy Failures

Why Conventional Mitigation Approaches Cannot Succeed

Howard A. Latin
Rutgers University, USA

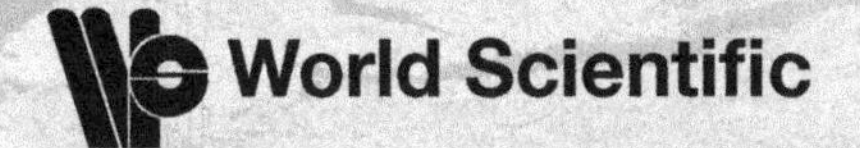

NEW JERSEY • LONDON • SINGAPORE • BEIJING • SHANGHAI • HONG KONG • TAIPEI • CHENNAI

Published by

World Scientific Publishing Co. Pte. Ltd.
5 Toh Tuck Link, Singapore 596224
USA office: 27 Warren Street, Suite 401-402, Hackensack, NJ 07601
UK office: 57 Shelton Street, Covent Garden, London WC2H 9HE

Library of Congress Cataloging-in-Publication Data
Latin, Howard A.
Climate change policy failures : why conventional mitigation approaches cannot succeed / by Howard A. Latin.
p. cm.
Includes bibliographical references and index.
ISBN-13: 978-9814355643 (hardcover)
ISBN-10: 981435564X (hardcover)
1. Climatic changes--Economic aspects. 2. Climatic changes--Government policy. 3. Greenhouse gases. 4. Climate change mitigation. I. Title.
QC903.L377 2011
363.738'74561--dc23

2011037451

British Library Cataloguing-in-Publication Data
A catalogue record for this book is available from the British Library.

In-house Editor: Wanda Tan

Typeset by Stallion Press
Email: enquiries@stallionpress.com

Printed in Singapore.

This book is dedicated to
Bernadette, Molly, Boozle, and Aunt Bunny

Contents

Acknowledgments

I have been assisted in writing this book by the ideas, comments, and criticisms of many people in the climate change, sustainable development, renewable energy, ecological economics, and environmental law fields. I can only name a small selection of these helpful academics and climate experts: James Hansen, William H. Rodgers, Jr., Douglas Kysar, Leslie McAllister, David Driesen, Daniel Farber, John D. Sterman, Michael Garrard, Robert Stavins, Lisa Heinzerling, Frank Ackerman, Richard Reibstein, Navroz Dubash, Carolyn Fischer, Bruce Pardy, and my colleagues, John Leubsdorf, Louis Raveson, and Paul Axel-Lute.

I greatly appreciate the editorial suggestions and information I received from my research assistant, Pamela Bingcang, and I also thank several years of Rutgers Law School students in my global climate change courses for their useful comments and citations.

The Social Science Research Network (www.ssrn.com) has provided access to thousands of scholarly papers on various dimensions of climate change. The International Institute for Sustainable Development (www.iisd.org) maintains the invaluable Climate-L email list, among others, and the Earth Negotiations Bulletin email list that distributes summaries of international negotiations on many relevant subjects. I have also benefited from the envlawprofessors email list, maintained by Professor John Bonine of the University of Oregon Law School, which has distributed thousands of comments and criticisms on diverse climate change issues.

I must gratefully thank Dr. Lim Tai Wei, Ms. Sandhya Venkatesh, Ms. Wanda Tan Hui Ping, Ms. Eliana Sidharta, Christine Yeung, and the other staff of the World Scientific Publishing Company for their assistance in bringing this book to fruition.

I would also like to express my great appreciation to Gary Dorin and Debra McPherson for their participation as board members of EcoVitality.org and for their efforts to help implement biodiversity conservation and sustainable development projects in comparatively poor countries.

Finally, not least but most, I must thank my wife, Bernadette Latin, for her constant encouragement and support, her many thoughtful suggestions, and her exceptional editorial assistance. She often understands my thoughts as well as I do and sometimes expresses them better.

CHAPTER I

Introduction: Challenging the Consensus

"Global climate change is the 'defining challenge of our era'."

UN Secretary-General Ban Ki-moon[1]

"We cannot afford more of the same timid politics when the future of our planet is at stake. Global warming is not a someday problem, it is now."

President Barack Obama[2]

In September 2009, the National Oceanic and Atmospheric Administration (NOAA) reported that "worldwide sea surface temperatures [were] the warmest at least since 1880, when such records were first systematically compiled."[3] In January 2010, the National Aeronautics and Space Administration (NASA) released data on Earth's surface temperatures showing that the decade ending in 2009 was "the warmest on record."[4] The hottest year was 2005; the second-hottest was 2009; and the "other hottest recorded years have all occurred since 1998."[5] In January 2011, NASA and NOAA announced jointly that the latest temperature readings show 2010 was tied with 2005 for the hottest year in recorded human history.[6] Annual and regional variations in surface temperatures are inevitable, but the trend toward greater global warming and climate change is unarguable from a scientific perspective.[7]

Recent research on climate change risks supports predictions of sea level rises, torrential rains and flooding, water scarcity and droughts, crop failures and famines, melting glaciers and ice sheets, severe heatwaves and wildfires, stronger hurricanes and extreme weather conditions, innumerable species extinctions, spreading tropical diseases, and other looming catastrophes.[8] These harms are the closest things we have experienced to the ten plagues of the Old Testament, except that climate change is occurring on a much broader scale and will last for a much longer time. And we are doing this to ourselves: pollution from millions of human activities has been damaging worldwide climate conditions because, despite all the dire warnings, people continue to discharge more heat-trapping gases into the atmosphere every year. The dimensions and timing of climate-related risks can seldom be quantified with scientific certainty; but if the world's leaders are unable to create intelligent, affordable strategies to overcome climate change dangers, all of these terrible predictions are very likely to come true in the near future.

Many publications have described the human hardships and ecological degradation expected from climate change.[9] The central purpose of this book, in contrast, is to challenge climate-policy *mistakes* that are leading to the adoption of misguided preventive programs with no chance of achieving genuine climate change progress. Whether as a result of ignorance, wishful thinking, self-serving rationalizations, or bad advice, most of the world's policymakers do not understand how to respond to growing climate change perils; and most established climate experts have been advising political leaders to choose feeble emissions-reduction approaches that cannot yield significant climate benefits. If we continue down the present ill-fated climate-policy trajectory, the program funding, personnel efforts, and precious time wasted on failing mitigation programs are sure to preclude or impede more sensible precautionary strategies.

When I was an unruly boy with a backpack full of alibis, my mother's favorite rejoinder was "the road to hell is paved with good intentions." This adage appears directly applicable to what is happening now in the realm of climate policy. Many leaders and expert advisors from developed nations advocate well-intentioned but unrealistic emissions-reduction measures that cannot improve the atmospheric conditions responsible for causing global climate change. These ineffectual climate-policy measures also cannot gain the essential cooperation of the major greenhouse-gas-polluting developing nations. Sincere concern without sufficient understanding is leading to the acceptance of self-defeating climate policies that will allow climate change risks to become steadily worse over the coming decades and centuries, despite the well-intentioned but fruitless expenditure of many billions of dollars on mitigation efforts that cannot succeed.

Many politicians and climate experts from developed nations are supporting a consensus strategy that relies on national or international emissions-reduction commitments to decrease greenhouse gas (GHG) discharges. These programs would gradually tighten the GHG emissions-reduction percentage rates over several decades, with limited cutbacks during the first 10 to 30 years followed by more ambitious pollution restrictions imposed by 2050. Most consensus mitigation proposals would also rely on cap-and-trade mechanisms and carbon offsets to reach GHG emissions-reduction targets by setting an arguably more efficient "price on carbon" and using market-system forces to create economic incentives for reducing greenhouse gas discharges.

Under these gradually more restrictive pollution controls, stringent GHG emissions-reduction targets will be deferred until several decades into the future. This deliberate delay is convenient for present-day politicians who can avoid making hard choices with high costs and unpleasant consequences; and it is also convenient for polluting businesses that can avoid implementing costly GHG clean-up measures for many years. But a multi-decade emissions-reduction delay is not convenient for the billions of people who are vulnerable to increasing climate change harms.

Here are brief summaries describing ten consensus GHG emissions-reduction programs or proposals included in recent American legal initiatives and international law agreements:

The Obama Campaign Proposal: During his presidential campaign, Barack Obama pledged that the United States would take the lead in cutting GHG emissions. He promised to impose "strong annual targets that set us on a course to reduce emissions to their 1990 levels by 2020 and to reduce the emissions 80 percent below 1990 levels by 2050."[10]

The Obama-for-President campaign website summarized their plan for relying on a cap-and-trade system with annually shrinking GHG limits that would impose graduated pollution cutbacks every year on the way to the ambitious 80% emissions-reduction target four decades from now.[11]

The Obama Administration Promises after the Copenhagen Negotiations: At the Copenhagen Conference of the Parties (COP 15) to the UN Framework Convention on Climate Change (UNFCCC)[12] held in December 2009, President Obama pledged that the US would undertake the GHG emissions reductions specified in the Waxman–Markey Bill adopted by the House of Representatives, presuming later Senate approval.[13] Following this climate-policy announcement, in January 2010 the President's Special Envoy for Climate Change, Todd Stern, submitted to the UNFCCC Secretariat an American commitment to reduce GHG discharges "in the range of 17 percent" in comparison to 2005 GHG pollution levels "in conformity with anticipated US energy and climate legislation," and Mr. Stern also noted that the "pathway set forth in pending legislation would entail a 30 percent reduction in 2025 and a 42 percent reduction in 2030, in line with the goal to reduce emissions 83 percent after 2050."[14] This "pending legislation" has now been rejected by the US Senate, and consequently the percentage emissions-reduction targets promised by President Obama and Mr. Stern have not been adopted into US regulatory law and evidently will not be realized in the foreseeable future.[15]

The Waxman–Markey Bill: In March 2009, Representatives Henry Waxman and Edward Markey introduced "The American Clean Energy and Security Act of 2009" (the ACES Bill), which included provisions to promote a "clean energy economy," to expand the use of "renewable energy sources," and to establish:

> a market-based program for reducing global warming pollution from electric utilities, oil companies, large industrial sources, and other covered entities that collectively are responsible for 85% of US global warming emissions. Under this program, covered entities

must have tradable federal permits, called "allowances," for each ton of pollution emitted into the atmosphere. Entities that emit less than 25,000 tons per year of CO_2 equivalent are not covered by this program. The program reduces the number of available allowances issued each year to ensure that aggregate emissions from the covered entities are reduced by 3% below 2005 levels in 2012, 20% below 2005 levels in 2020, 42% below 2005 levels in 2030, and 83% below 2005 levels in 2050.[16]

The Waxman–Markey Bill was approved by the House of Representatives after numerous weakening amendments — such as changing the 2020 GHG emissions-reduction target from 20% to 17% of 2005 discharges. The House Committee on Energy and Commerce approved the bill on a largely partisan vote of 33 to 25, and then in June 2009 the full House accepted a heavily-revised version by a vote of 219 to 212.[17] These weak political endorsements showed a lack of broad congressional support for climate change legislation,[18] but they nevertheless made the Waxman–Markey Bill the first climate change legislation to reach this stage in the American political process.[19] The approved bill fully incorporated the consensus graduated emissions-reduction strategy[20] that this book contends cannot be successful at reducing climate change risks or achieving meaningful climate benefits.

The Kerry–Lieberman Bill: The "American Power Act of 2010"[21] (APA) submitted by Senators John Kerry and Joseph Lieberman was co-sponsored by Senator John Graham, a Republican from South Carolina, but he withdrew his support from any climate change legislation because of his opposition to the Democratic immigration policy.[22] This unilateral repudiation showed limited political support in the Senate for climate and energy legislative programs, and no other Republicans came forward in 2010 to support this 987-page bill replete with special-interest payoffs and loopholes. Because the Kerry–Lieberman Bill failed to attract a majority of the Senate, to say nothing of the 60% vote required for cloture,[23] there is little point in analyzing its many complicated provisions and deciding whether they would have been able to help reduce climate change risks, which I very strongly doubt.

The Kerry–Lieberman Bill subscribed to nearly the same emissions-reduction policies and pollution cutback targets as those in the Waxman–Markey Bill — imposing a timetable for cutting GHG emissions from major sources by 4.75% below 2005 discharge levels by 2013, 17% below 2005 discharge levels by 2020, 42% below 2005 levels by 2030, and 83% below 2005 pollution levels by 2050.[24] This Senate bill also included the equivalent of a cap-and-trade mechanism restricted to energy-industry GHG sources, as well as various agricultural and forestry carbon offset provisions meant to achieve the required emissions reductions in a supposedly more efficient or less costly manner.[25] If we overlook the billions of dollars this bill would have allocated to new nuclear energy plants, ostensibly "clean" coal-burning power plants, more offshore oil drilling and natural gas development, and the imposition of market price "collars" to minimize the amount of clean-up costs that could be imposed on "dirty" GHG-polluting businesses, the heart of the Kerry–Lieberman Bill consisted of the consensus multi-decade, graduated emissions-reduction strategy that this book maintains cannot reduce the severity or frequency of climate change dangers.

The California Clean Car Standards and Federal Fuel-Efficiency Regulations: In one of the first state initiatives to address climate change, the California Clean Car regulations of 2006 required a 30% reduction in GHG emissions from new motor vehicles by 2016.[26] In September 2009, the Obama Administration instructed the US Environmental Protection Agency (EPA) to promulgate similar nationwide fuel-efficiency regulations applicable by 2016 that would impose about a 30% GHG reduction target and provide national uniformity in the GHG emissions-reduction standards that car and light-truck manufacturers must meet for new motor vehicle discharges.[27] In July 2011, President Obama pledged to increase automotive fuel efficiency standards gradually to a fleet average of 54.5 miles per gallon by 2025, but it remains to be seen whether this substantial improvement can be implemented over the next 15 years.[28]

The RGGI Program: Nearly a dozen Eastern US states are participating in the Regional Greenhouse Gas Initiative (RGGI), which is "the first mandatory, market-based effort in the United States to reduce greenhouse gas emissions. Ten Northeastern and Mid-Atlantic states will cap and then reduce CO_2 emissions from the power sector 10% by 2018."[29] To achieve the required GHG reductions in as inexpensive a manner as feasible, the RGGI program set up the first carbon-based cap-and-trade system in the United States.

The WCI Program: Following the RGGI model, the Western Climate Initiative (WCI) multi-state agreement provides that: "The cap-and-trade design is an important element of a comprehensive regional effort by the governors and premiers of seven US states and four Canadian provinces to promote environmental sustainability and economic growth by reducing greenhouse gas emissions by 15% below 2005 levels by 2020."[30]

The Kyoto Protocol: On an international plane, the Kyoto Protocol obligates developed nations (Annex I states-parties) to reduce their GHG emissions on average about 5% below 1990 levels, while the Protocol imposes no specific emissions-reduction commitments on developing countries.[31] Years of inconclusive international negotiations have failed to determine how, or how much, to increase the Kyoto Protocol emissions-reduction requirements after the agreement's initial commitments expire in 2012, but all of the various proposals remain based on the consensus plan for gradually cutting GHG discharges from Annex I developed states over the next several decades.[32]

European Union Initiatives: The European Union nations have devoted several years to debating what their next emissions-reduction goal should be after the initial Kyoto Protocol targets expire in 2012, despite the failure of most EU nations to attain the minimal Kyoto restrictions until the business contractions caused by the recent worldwide recession.[33] In 2008, the EU adopted a "20–20–20" policy that would require a 20% reduction from 1990 GHG emissions levels by 2020 and

20% reliance on renewable energy sources by the same year.[34] The EU plans to rely mainly on the third phase of the Emissions Trading Scheme (ETS), the world's largest cap-and-trade system, to achieve its projected 20% emissions cutback target.[35] The EU Commission has been considering the practicality of imposing a more stringent 30% emissions-reduction target by 2020 if other developed nations make comparable commitments.[36]

United Kingdom Plans: The "2008 Climate Change Act requires British dischargers to reduce emissions by at least 34 percent by 2020 from the levels reached in 1990."[37] In addition to this law and the EU programs, the Chair of the United Kingdom Committee on Climate Change, which was created by the government to provide input on climate change policies, recently reported that the UK "must slash global-warming gases by 80 percent rather than the current 60 percent target to avert rising temperatures that may have a 'major and increasing' impact on humans and the environment."[38] The advisory panel "must also set carbon 'budgets' for government, outlining the cuts that need to be made in successive five-year periods."[39]

Then in May 2011, Chris Huhne, the British energy and climate secretary, announced that the "carbon budget — a 50% cut averaged across the years 2023 to 2027, compared with 1990 levels — would be enshrined in law."[40] This planned GHG cut is the most stringent emissions-reduction commitment yet made by any developed nation, but it is still only a promise deferred until the future, not an implementation reality.

Most major environmental groups working in the arena of climate policy have also supported the consensus graduated GHG emissions-reduction approach. Before the recent congressional bills were introduced, the Natural Resources Defense Council advocated emissions cuts of "25 percent by 2020 and 80 percent by 2050."[41] The Environmental Defense Fund asked members to support the Lieberman–Warner Climate Security Act of 2008, which they claimed would have cut GHG pollution "almost 20% below current levels by 2020, while setting us on the path to the 80% emissions reductions scientists say we need by mid-century."[42] And the Union of Concerned Scientists proposed that the world's industrialized nations "reduce their emissions an average of 70 to 80 percent below 2000 levels by 2050."[43] These progressively stricter emissions-reduction proposals from non-governmental organizations (NGOs) explicitly relied on the consensus graduated GHG pollution control strategy that I am challenging.

My goal here is to describe the similarity of nearly all proposed emissions-reduction programs and to explain why they have been, or will be, undermined by the same fundamental climate-policy mistakes. These programs and proposals all share the graduated multi-decade emissions-reduction strategy and almost all would employ cap-and-trade systems to "put a market price" on GHG discharges. Although the selected cutbacks and target dates would differ to a limited degree, the emissions-reduction strategy and its major weaknesses are essentially the same for all of these climate change mitigation efforts.

As emphasized in Chapter II of this book, the consensus approach will prove "too little, too late" by deferring crucial GHG reductions too far into the future. These emissions-reduction programs would consistently be *back-loaded*, in the sense that major GHG pollution decreases will not be imposed until the last decades of the target schedule or later. None of the consensus proposals would require annual GHG emissions cutbacks of even 50% until 2050 and thereafter. Before then, weak interim emissions-reduction targets would allow huge amounts of *residual* GHG discharges — all greenhouse gas discharges authorized within the annual emissions-reduction targets or caps — which are certain to compound the already-too-high atmospheric GHG concentration while allowing the greenhouse effect, global warming, and climate change to become steadily worse.

The most common greenhouse gas, carbon dioxide, is highly persistent and may remain in the atmosphere for centuries or longer.[44] This persistence means that climate change hazards exacerbated by the next several decades of large residual GHG discharges cannot later be undone by the stricter emissions-reduction targets that supposedly will be imposed after 2050. The cumulative build-up of residual GHGs in the atmosphere allowed by the consensus emissions-reduction programs, combined with the long-term persistence of CO_2, will prevent these feeble consensus mitigation efforts from effectively stabilizing or decreasing the atmospheric GHG concentration that causes global climate change.[45]

At best, the consensus emissions-reduction programs will only *slow the growth* of the atmospheric GHG concentration and related climate change risks to a minimal extent. Chapter II shows that the required GHG pollution cuts under the consensus emissions-reduction programs will only "*reduce the increases*" in the atmospheric GHG concentration. As a result of these undesirable cumulative increases, the consensus emissions-reduction programs would probably create an illusion of climate change mitigation progress without actually reducing the greenhouse effect or attaining any discernable climate benefits. The climate-policy mistake of authorizing vast amounts of persistent residual GHG discharges that will increase heat-trapping atmospheric GHG levels for the next four decades and beyond is certain to undermine any potential benefits from the consensus mitigation programs, wasting irreplaceable time and resources that could be used instead to implement more promising climate-control efforts.

I want to make very clear my contention that the recent congressional bills, EU promises, and Obama Administration proposals, if they are ever implemented, would guarantee that climate change dangers will continue to grow worse. The consensus emissions-reduction programs cannot achieve any significant climate benefits because they do not go nearly far enough fast enough in confronting the cumulative effects of degraded climate conditions and increasing GHG pollution levels. These are much more fundamental criticisms than the frequent environmentalist complaint that the consensus emissions-reduction proposals should be stronger. I am arguing that even if the proposed consensus greenhouse gas reduction programs work as well as they realistically can, they will still be wholly inadequate to curtail ongoing climate degradation with associated global harms.

Chapter III evaluates cap-and-trade mechanisms, carbon offsets, and carbon taxes among the economic incentives schemes proposed by the Waxman–Markey and Kerry–Lieberman Bills, the Obama Administration proposals, and other market-forces proponents. The central theme of this assessment is that the market-based approaches are at least equally prone to allow constant increases in the atmospheric GHG concentration resulting from persistent residual GHG discharges. If this assessment is correct, the highly-praised market-based mechanisms will prove no more effective than "command and control" regulation in restraining the ongoing degradation of climate conditions. Allowing large volumes of persistent residual GHGs to be discharged into the air under the guise of tradable GHG allowances or carbon offset credits will be just as likely as direct regulation to cause the atmospheric GHG concentration to increase while climate change risks become progressively more hazardous.

Chapter IV examines the goals, characteristics, and prospects of continuing reliance on international negotiations to reach a global solution for climate change problems. This chapter contends that the consensus emissions-reduction approaches sponsored by many policymakers and expert advisors in "Northern" developed countries will be rejected by the large GHG-polluting "Southern" developing nations, which regard improving their economic and social welfare as their highest priority. The consensus emissions-reduction programs of the developed nations do not enable the less affluent states to obtain *what they want most* — greater economic prosperity. Instead, the consensus Northern mitigation programs would be certain to create unwelcome barriers to development in many comparatively poor countries.

An effective international climate policy must be able to do *both* concurrently: It must promote greater economic and social welfare in developing nations by disseminating GHG-free technologies and foreign assistance funding that will enable increased development without further damaging the climate. In effect, a successful international climate policy must be able to reduce GHG discharges leading to climate change harms while it increases "clean" development in many nations that regard economic growth as their highest priority. International negotiations based on the imposition of consensus emissions-reduction programs that require substantial GHG pollution cutbacks by all nations and ignore what most developing states want intensely are bound to prove politically untenable and wholly ineffective.

Chapter V presents my climate-policy recommendations with a reasonable level of detail. I contend that resolving climate-policy problems, collective-action problems, and development problems will require shifting our primary mitigation efforts and investments from a GHG emissions-reduction strategy to a *GHG-free replacement-technology strategy*. The best way and probably the only way to reduce GHG discharges *enough* to avoid escalating climate change dangers is to eliminate the major sources of GHG pollution as rapidly as feasible in as many sectors and regions as feasible by replacing them with comparable GHG-free technologies, processes, and methods. This climate-policy approach is sometimes called "decarbonization."

Because CO_2 will persist in the atmosphere for centuries or longer, we must begin to replace "dirty" GHG-polluting sources with "clean"[46] replacement technologies and processes as promptly as possible to the greatest extent feasible. A "clean" replacement-technology approach is the only realistic way to eliminate the large quantities of persistent residual GHG discharges that are the fatal flaw of the consensus emissions-reduction programs, and it is also the only viable way that affluent nations can meet the economic needs and prosperity goals of the developing countries without continuing to degrade global climate conditions.

We cannot phase out all GHG discharges, which would be impossible in many poor regions where the highest priority is expanding economic prosperity,[47] but a GHG-free replacement-technology approach would likely be the most effective way to remove greenhouse gases that can feasibly be eliminated. GHG pollution control regulatory programs or economic incentive instruments, such as cap-and-trade systems and carbon taxes, cannot succeed in overcoming major climate change problems until there are clean alternative technologies that can perform the essential industrial and consumption functions now achieved by fossil fuel combustion and other traditional GHG sources. Instead of retaining "dirty" technologies that put out large amounts of residual GHGs, and trying to decrease the climate change risks by making these GHG sources a little "more efficient" or somewhat "less polluting," we should put our primary efforts and investments into creating replacement technologies capable of accomplishing the same kinds of vital economic and social tasks without continuing to damage Earth's climate.

Almost everyone in the climate field acknowledges the need for GHG-free replacement technologies *someday*, especially in the primary energy sector, but the required expenditures, technological innovations, and widespread deployments are usually regarded as something we should try to achieve sometime in the future when GHG emissions-reduction technologies and alternative energy sources have become more certain and less costly. In the meantime, most policymakers and climate experts go on debating how large or how small a percentage cutback of business-as-usual GHG discharges should be mandated by the consensus emissions-reduction programs they happen to favor. Yet, the ill-conceived consensus emissions-reduction programs, including those relying on cap-and-trade systems, will waste irreplaceable time and resources while allowing climate change risks to become steadily more severe. Now, today, not tomorrow, we need to put our greatest efforts and financing into eliminating, or greatly reducing, society's reliance on fossil fuels, other carbon combustion methods, carbon-based industrial chemical processes, energy-wasting vehicles and buildings, and other GHG-generating activities that are steadily causing climate change to become more harmful and longer lasting.

Orchestrating an effective mitigation plan to overcome difficult climate problems and collective-action problems requires implementation of several overlapping, mutually-supportive institutions described in Chapter V. It is important to recognize that no single mitigation measure or remedial approach is likely to succeed if it is pursued as the only "solution" on which society must depend. The development and deployment of GHG-free replacement technologies is the lynchpin of my proposals, but we must find reliable ways to pay for these alternative technologies and methods, we must have a way to deal with GHG sources that do not now have feasible clean alternatives, and we need to employ as many ways as possible to persuade major GHG sources to adopt more climate-friendly practices.

Policymakers should recognize that the numerical cutback targets of the consensus GHG emissions-reduction programs cited in this chapter are speculative numbers imagined for a single moment in time — economists call this a *static* perspective — although world-wide development activities and climate conditions are changing *dynamically* every year. If the GHG emissions from an activity are reduced by 50%, cutting greenhouse gas pollution in half for each unit of a given activity, while the number of output units doubles over time, there will be no net reduction in GHG discharges. This is a vital consideration because world population is expected to grow to about nine billion people by 2050,[48] with the great majority of this growth taking place in developing countries whose people have rising aspirations for improved economic and social welfare. Meeting these rising aspirations will be impossible without commensurately increasing the GHG emissions coming from rapid economic growth if these people follow the same "dirty" historical development patterns that the Northern states created. In effect, we will have to run as hard as we can just to try to stand still in terms of stabilizing atmospheric GHG levels, and even that may prove impossible as world population and development goals keep increasing.

Viewed in this light, the consensus mitigation measures supposedly sufficient to achieve an 80% GHG reduction by 2050 will no longer be good enough in 2051 and future years. We cannot ever stop reducing GHG pollution discharges if we intend to attain any of these arbitrary numerical emissions-reduction targets. Perhaps one moment will occur when the atmospheric GHG level is counter-balanced by the natural and human-made processes ("sinks") that remove GHGs from the air — this is called a "stabilization point" in the climate literature[49] — but the next moment someone somewhere will decide to go for a long drive, take a distant vacation, or build a new factory, and the atmospheric GHG stability will again become unbalanced.

GHG emissions-reduction programs or other mitigation efforts must greatly improve their performance to meet the increasing GHG discharges from more economic development activities benefiting more people in more places. In reality, mitigation programs cannot possibly keep up with this dynamic growth and commensurate social demands unless almost all GHG-based technologies and processes can eventually be replaced by equivalent clean GHG-free production and consumption methods that could be scaled upwards to accommodate expected population growth and greater economic growth while still remaining GHG-free.

It is also important to acknowledge that many GHG-producing activities cannot feasibly be eliminated in the near future. Until the entire world is electrified using affordable alternative energy technologies, for example, billions of poor people will go on burning wood or charcoal for heating and food preparation purposes. We cannot expect these people to behave otherwise until low-cost GHG-free replacement heating technologies are available and affordable that can achieve the necessary subsistence functions without degrading climate conditions. Harmful GHG emissions cannot be completely eliminated from agriculture, transportation, existing buildings, and many other common pollution sources within a timeframe of several decades, which means that these continuing sources of GHGs must be accommodated within whatever long-term emissions-reduction plan or timetable is chosen.

Many recreational, cultural, or lifestyle activities attract people with strong preferences who will not give up their cherished behaviors even when no GHG-free alternatives exist. Some of these "dirty" but satisfying activities include NASCAR auto racing, sport fishing, recreational flying, aesthetic wood fires and campfires, mountain climbing in remote ranges, vacations to Disney World or beach resorts, beef consumption, whisky distilling, and space explorations, to name only a few among millions of activities that now create substantial GHG pollution in the aggregate. Because carbon dioxide, the most common GHG, is highly persistent, emissions from diverse consumption activities must be included in any long-term mitigation plan meant to reach strict emissions-reduction targets in near-future decades.

In recognition of the difficult impediments to overcoming global climate change and our inability to eliminate all GHG emissions, I see only one emissions-reduction target that appears suitable for developed nations: We need to eliminate as much GHG pollution as feasible as soon as feasible in as many contexts as feasible. In other words, the goal should be to eliminate GHG pollution to the greatest feasible extent because we cannot afford to cut out all GHG emissions in a realistic timeframe and the demands for greater GHG pollution will keep growing as world population and development increase. Aiming for weak back-loaded GHG pollution limits, such as those included in the consensus emissions-reduction programs cited in this chapter, cannot be sufficient to combat dynamically increasing climate change problems, and even the standard of eliminating all GHG emissions to the greatest extent feasible will not be sufficient unless the large GHG-polluting developing countries, such as China and India, can be persuaded to join global mitigation efforts.

If climate policymakers accept the conclusion that developed nations will have to make extreme reductions in GHG emissions, coming as close to zero discharges as they feasibly can despite the existence of many activities that will continue to generate some GHG pollution, perhaps these leaders and their expert advisors will stop debating arbitrary percentage-reduction targets and will agree that deploying GHG-free replacement technologies and methods must be the primary strategy for meeting GHG mitigation goals. In the words of scholars who have reached similar climate-policy conclusions, society must create "transformative technologies" able to replace GHG pollution sources that are leading to growing climate change perils.[50]

If we dither for decades and waste many billions of dollars attempting to make modest reductions in GHG emissions that will yield no substantial climate benefits, while more and more persistent residual discharges are pumped into the air every year, we will still eventually have to adopt a "clean" replacement-technology strategy that will likely be more difficult and expensive to achieve than if we start moving toward a GHG-free or very-low-GHG society now. As world population, economic development pressures, and energy needs keep expanding, we will have to replace as many harmful GHG sources as feasible rather than continuing along the same "dirty" development path while adopting a patchwork-quilt of futile emissions-reduction band-aids.

Whether or not readers agree with my focus on the critical need for the development and dissemination of GHG-free replacement technologies, the most important contribution of this book is to show that the consensus emissions-reduction approaches now advocated by most climate policymakers, expert advisors, and prominent environmental groups cannot be effective at reducing climate change dangers due to the cumulative amount and persistence of the residual GHG discharges that will be pumped into the atmosphere during the next several decades under these ineffectual climate policies. We must discard wishful thinking based on consensus emissions-reduction programs that are *convenient* for governments now, but cannot succeed in restraining, much less overcoming, increasing climate change risks every year in the future.

The goal of this book is to offer persuasive explanations and criticisms of current climate-policy failures, including the consensus emissions-reduction programs, which will achieve no tangible climate progress while they undermine potentially more effective solutions. The burden of persuasion should be placed on the advocates of back-loaded GHG emissions-reduction measures to show how these programs could produce worthwhile climate change improvements commensurate with their large residual GHG emissions, long delays, and high implementation costs; while the burden of persuasion should fall on me to show that the arguments of consensus emissions-reduction advocates are indisputably wrong.

Whether or not readers agree with my focus on the [illegible] [illegible] development and [illegible] [illegible] [illegible] [illegible] the most important [illegible] of the book is [illegible] that greenhouse gas emissions reduction approaches [illegible] various climate policy [illegible] and [illegible] environmental groups, cannot be [illegible] [illegible] the [illegible] amount and persistence of the [illegible] GHG discharges that will be pumped into the atmosphere during the next several decades under the current climate policies. We must [illegible] based on [illegible] emission-reduction [illegible] for governments [illegible] [illegible] to [illegible] much less [illegible] [illegible] change [illegible] in the future.

The goal of this book is to offer persuasive explanations and [illegible] on current climate policy [illegible], including [illegible] [illegible] why [illegible] [illegible] effective [illegible] [illegible] should [illegible] [illegible] to show how these [illegible] [illegible] costs, [illegible] arguments [illegible] emissions reduction advocates are [illegible].

CHAPTER II

"Reducing the Increases" in the Atmospheric GHG Concentration

As UN Secretary-General, Ban Ki-moon, recently observed: "The bottom line is that our climate is changing fast, and the world has been too slow in response."[51] Yet, the major GHG-polluting nations have not come close to agreeing on realistic policies for avoiding or mitigating climate change risks. Conflicting national interests and priorities have undermined many climate change proposals, but of equal importance most political leaders do not realize that they lack a clear understanding of climate-related issues; and therefore, they are bound to continue making climate-policy mistakes with tragic consequences.

In basic terms, the greenhouse effect, global warming, and climate change are caused by the retention of excess heat-trapping gases in the atmosphere.[52] Increasing the amount of GHGs in the air will worsen the greenhouse effect by trapping more heat, which means that the *concentration of greenhouse gases in the atmosphere* is the primary factor determining the extent of climate change risks.[53] Readers might suppose that the most important climate policies would focus on stabilizing and then reducing the atmospheric GHG concentration that is causing the greenhouse effect. This is what a sensible climate policy could do and should do. Yet, nearly all climate change mitigation plans around the world, including all of the consensus emissions-reduction initiatives summarized in Chapter I, focus instead on reducing current or projected GHG discharges by selected percentage rates applied over several decades. The conventional mitigation programs almost always ignore the cumulative impacts of persistent residual discharges on the atmospheric GHG concentration, which is a crucial mistake.

These two kinds of pollution control targets — reducing the annual GHG discharges by a designated percentage rate, or decreasing the atmospheric GHG concentration — may seem to be equivalent and many people treat them as the same, but they definitely are not the same. Under a GHG emissions-reduction approach, the claimed "reductions" come from comparisons with the amount of GHGs that would be discharged by pollution sources if no regulatory controls or caps were imposed, and not from comparisons with the amount of GHGs *already in the air*. In the literature on climate change, this do-nothing or no-regulation condition is described as the business-as-usual (BAU) scenario or trajectory.[54] The consensus emissions-reduction programs are intended to cut BAU discharges by a selected percentage rate over a multi-decade period, while climate policymakers ordinarily do not even consider the ramifications of these emissions-reduction efforts for the atmospheric GHG concentration.

A major cause of climate-policy confusion and mistakes is the presence of two different background levels or *baselines* against which changes in climate conditions can be measured. One baseline compares the amount of GHG discharges eliminated by an emissions-reduction program against the BAU pollution volume for a given year. This BAU baseline comparison addresses how much GHG pollution will be cut by a selected mitigation program in comparison to BAU discharges with no cutbacks whatever. In contrast, the alternative baseline compares the atmospheric GHG concentration in a given year against the atmospheric concentration in a prior year or projected future year. This baseline comparison addresses a mitigation program's annual impact on the aggregate atmospheric GHG concentration that causes the greenhouse effect in comparison to rising, declining, or stable atmospheric concentrations in other years.

Consider the Regional Greenhouse Gas Initiative (RGGI), which has chosen a target of eliminating 10% of GHG emissions from power plants in the eastern US by 2018.[55] If this program is viewed from the BAU discharges baseline, it appears to accomplish some positive results by cutting eastern power plant GHG discharges 10% over the course of a decade. Using a BAU baseline, any emissions-reduction program that *cuts any amount* of GHG pollution over any duration will appear to be at least a marginally worthwhile achievement when compared against GHG discharges with no restrictions at all placed on them. Under this BAU baseline, it would be easy for concerned people to conclude mistakenly that all GHG emissions-reduction efforts are bound to do "some good" and also to believe the erroneous "every little bit helps" idealistic but unrealistic perspective.

In contrast, if the atmospheric GHG concentration baseline is adopted, as it should be, the RGGI program would be characterized as cutting 10% of BAU discharges from regulated power plants while allowing the remaining or residual 90% of business-as-usual GHGs to be released into the air each year. Under the RGGI program, the power plants will be allowed to continue discharging 90% of the GHG emissions that they would have discharged absent any pollution regulation. Praise does not seem warranted for a GHG emissions-reduction program that authorizes the continuation of 90% of annual BAU discharges from regulated power plants and consequently will allow millions of tons of additional GHG emissions to increase the aggregate atmospheric GHG concentration in the next decade. The crucial point is that by authorizing large residual GHG discharges from the regulated power plants, which would increase the atmospheric GHG concentration each year, the RGGI mitigation initiative will consistently allow climate change conditions to become more hazardous rather than safer.

The remaining or residual GHG discharges — the volume of greenhouse gas pollution that the RGGI power plants will be allowed to put out above the 10% emissions-reduction target — would constantly increase the atmospheric GHG concentration, causing more global warming and climate change. Applying the BAU baseline to the RGGI program makes it appear to be a minor mitigation achievement cutting power plant emissions by 10% annually. On the other hand, applying the atmospheric concentration baseline shows that the RGGI program would be a major contributor to ongoing climate degradation because it will allow 90% of harmful BAU discharges from eastern power plants to further contaminate the atmosphere every year.

Taking into account the financial and political capital, administrative resources, personnel efforts, and irreplaceable time invested in the RGGI emissions-reduction program, and thus diverted from more beneficial regulatory efforts, this emissions-reduction program should be viewed as a serious climate-policy mistake rather than a limited success. As long as the RGGI program authorizes huge residual GHG discharges that will increase the cumulative atmospheric GHG concentration, this mitigation program cannot produce tangible climate change benefits despite its benign intentions and supposedly "efficient" cap-and-trade mechanism.

In May 2011, New Jersey's newly-elected conservative Republican Governor, Chris Christie, announced that he will withdraw his state from the RGGI program because it is "an ineffective way to reduce carbon dioxide emissions."[56] Governor Christie complained: "RGGI does nothing more than tax electricity, tax our citizens, tax our businesses, with no discernible or measurable impact upon our environment."[57] This position is consistent with the views of many cap-and-trade critics, who contend that the measures "constitute a new form of taxation because they impose additional costs on electric utilities that are then passed on to customers."[58] While disparaging the RGGI, Governor Christie asserted that New Jersey is already reducing GHG emissions because it is "relying more on natural gas and less on coal to fill its energy needs."[59]

I do not agree with this hostility to climate regulation because the anti-tax ideologues are failing to consider the very high costs imposed on society by diverse climate change harms, which are implicit "taxes" imposed by "dirty" industrial processes relying on fossil fuel combustion. The vital lesson from New Jersey's RGGI withdrawal is that climate change mitigation programs need to show they are producing *significant benefits* in return for their substantial costs. I must reluctantly agree with Governor Christie that the RGGI has been a waste of irreplaceable time and resources without in any way achieving its precautionary goals. Unsurprisingly, Mr. Christie does not appear to be looking for more effective emissions-reduction methods, and substituting one harmful fossil fuel (natural gas) for an even more harmful fossil fuel (coal) is hardly a sufficient carbon mitigation strategy.

Another illustration of the importance of applying the atmospheric GHG concentration baseline, rather than the BAU baseline, involves the California Clean Car initiative and recently updated federal fuel-efficiency standards, which both require new motor vehicles to reduce their GHG emissions roughly 30% by 2016 and around 54% by 2025. Under a BAU baseline, these emissions-reduction initiatives may appear beneficial because they will eventually reduce GHG discharges from new vehicles by about one-third in 2016 and about one-half in 2025. In contrast, under the atmospheric GHG concentration baseline, these consensus mitigation measures should be evaluated based on how their large residual GHG discharges will affect the atmospheric concentration during succeeding years.

The California regulations and federal fuel-efficiency standards will do nothing to limit increased GHG discharges from a growing number of cars on the road,[60] from a growing number of vehicle-miles driven, or from GHG emissions resulting from more frequent traffic jams. The Clean Car regulations will only cut about 30% of BAU discharges from new vehicles while allowing the remaining *70% of GHG emissions* from new vehicles and 100% of emissions from existing (used) vehicles to be discharged into the atmosphere annually. By the regulatory target date of 2016, new motor vehicles in California and other states will have discharged millions more tons of residual GHGs, contributing to the continuing greenhouse gas build-up in the atmosphere.

If we assess the Clean Car programs using the BAU discharges baseline, they may appear "successful" in reducing GHG emissions from new vehicles by one-third. In contrast, these programs will authorize huge amounts of persistent residual GHG discharges that will be added to the atmosphere each year, consistently increasing the GHG concentration in the air. The California and federal regulations will allow more than twice as much GHG emissions to reach the atmosphere in comparison to the 30% reduction of BAU discharges these programs are meant to attain. If we contrast the atmospheric GHG concentration in 2006 with the expected concentration in 2016 after a decade of these Clean Car programs, the residual GHG emissions from millions of new vehicles are bound to make climate conditions worse rather than "cleaner."

As another illustration of using the atmospheric GHG concentration baseline, in July 2008 the leaders of the G8 developed countries, including Bush Administration officials, agreed to reduce annual GHG emissions 50% by 2050[61] without identifying any interim pollution control targets. An emissions-reduction program that cuts BAU discharges from a country or a company by 50%, such as the G8 leaders agreed to do, would allow the remaining 50% of GHG pollution to become residual GHG emissions that will increase the cumulative atmospheric GHG concentration and greenhouse effect. In effect, the BAU greenhouse gas volume may be *reduced* while the annual residual discharges continue to *increase* the atmospheric GHG concentration.

These examples show that reducing GHG emissions compared to BAU pollution levels while retaining large residual GHG discharges ordinarily will increase, not reduce, the aggregate atmospheric GHG concentration and related climate change dangers.

Consider the pattern of GHG discharges illustrated in Figure 1.

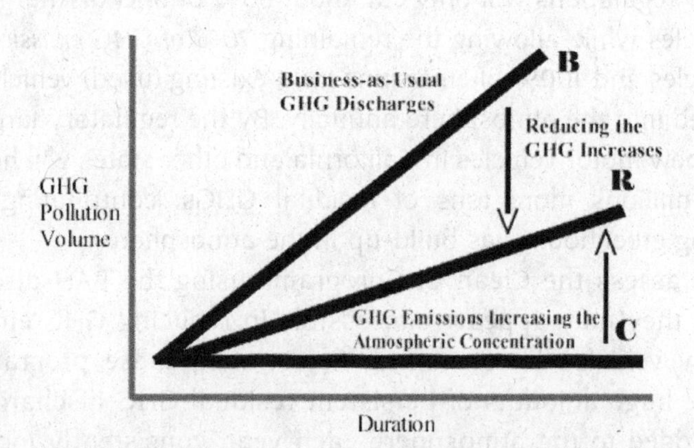

Figure 1: "Reducing the Increases" in the Atmospheric GHG Concentration

A GHG source generating high business-as-usual discharges at level B could cut its GHG pollution more than 50% to level R by adopting various mandatory or voluntary emissions-reduction measures. This is the percentage cutback process relied on by consensus emissions-reduction programs, including all consensus programs cited in Chapter I. At emissions level **R**, however, this pollution source would still be adding large amounts of residual GHGs to the existing atmospheric GHG concentration at level **C**. The GHG emissions reductions from level **B** to level **R** will nonetheless leave substantial residual discharges that would increase the atmospheric GHG concentration (level **R** added to level **C**) and climate change harms.

Focusing on *reducing* the annual BAU volume of GHG pollution rather than limiting the *increases* in the atmospheric GHG concentration is a serious climate-policy mistake that will destroy the effectiveness of emissions-reduction programs. Mitigation programs that reduce the annual BAU pollution level while they allow large residual GHG discharges to reach the air normally cannot stabilize or reduce the atmospheric greenhouse effect and climate change dangers because the new GHG discharges will *combine* with the GHGs already in the air to increase the cumulative GHG concentration.

In explaining the dynamic "stocks and flows" properties of the GHG concentration in the atmosphere, Professor John Sterman of MIT observed:

> Our mental models suggest that if we stop the growth of emissions, we will stop global warming, and if we cut emissions, we'll quickly return to a cooler climate. We tend to think that the output of a process should be correlated with — look like — its input. If greenhouse gas emissions are growing, we think, the climate will warm, and if we cut emissions, we imagine that the climate will cool. In systems with significant accumulations, however, such correlational reasoning does not hold. Rather, it's more like filling a bathtub. The amount of carbon dioxide in the atmosphere is like the level of water in a bathtub. The level grows as long as you pour more water in through the faucet than drains out. Right now, we pour about twice as much CO_2 into the atmospheric tub than is removed on net by natural processes.[62]

Although the annual GHG emissions volume has been *reduced* from level **B** to level **R** in Figure 1, the residual GHG discharges at level **R** will combine with the prior "stock" of GHGs at level **C** to *increase* the atmospheric GHG concentration. Limited emissions reductions from BAU pollution levels that allow large annual residual GHGs to combine with the already-too-high existing atmospheric GHG level cannot attain any progress in controlling climate change risks. The "reduced" GHG emissions stream will still be harmful, not helpful, because the new discharges will *combine* with the existing "stock" of GHGs in the air to raise the aggregate atmospheric concentration. As a consequence of this dynamic "stocks and flows" process, there is no rational basis for the widely-held view that every little bit of GHG pollution reduction will help to promote climate change improvement or to lessen climate change hazards.

For example, one environmental law professor who specializes in climate issues wrote: "Each unit of carbon dioxide (CO_2) not emitted is one unit that will not persist in the atmosphere for 100 years, and each reduction, no matter how small, reduces overall GHG concentrations in the atmosphere."[63] Yet, when residual GHG discharges are added to the current atmospheric GHG level, the cumulative atmospheric concentration will grow worse even if some of the BAU emissions have been decreased. When the author of this statement referred to "each reduction, no matter how small," he was using the BAU baseline, comparing reduced discharges against larger BAU discharges. He was not considering the cumulative impact of the remaining residual GHG emissions on the large "stock" of GHGs already in the air.

If the atmospheric GHG concentration baseline is employed instead, cutting BAU discharges by a "little bit" normally will not reduce the cumulative atmospheric GHG "stock" or the resulting greenhouse effect, and the claim that all minor BAU emissions reductions will help decrease climate change risks is simply wrong. Even "little bits" of mitigation have their costs, including diverting public attention from more important precautionary efforts. Climate change preventive programs will almost always have monetary costs and opportunity costs that may outweigh any asserted benefits of "little bits" of mitigation even if there is some reduction in the BAU discharge levels. The quote from the professor above does not even mention the costs and foregone opportunities of these "little bits" of mitigation, which is a rather serious omission.

Proponents of the "every little bit helps" fallacy ask consumers and businesses to decrease their "carbon footprints" in many ways on the rationale that any GHG reductions will commensurately reduce climate change risks. Yet, this mantra does not take into account the reality that even large GHG reductions by millions of consumers may not lead to positive climate change benefits due to the residual pollution from diverse activities that in the aggregate will *increase* the crucial atmospheric GHG concentration. As an example, let us consider the behavioral changes shown in Table 1 that could lead to "reducing the increases" from automobile purchases, thereby achieving little or no climate change progress.

Table 1: Assessment of 2010 Hybrid Vehicle Models by Miles Per Gallon (MPG) and GHG Discharges

Vehicle Models	Type	Highway MPG	City MPG	Annual Tons of GHGs Emitted
Toyota Camry	4-cylinder auto	33	22	7.2
Toyota Camry	Hybrid	34	33	5.5
Toyota Prius	Hybrid	51	48	3.8
Ford Fusion	All-Wheel-Drive	24	17	9.8
Ford Fusion	Hybrid	41	36	4.8
Ford Escape	4-Wheel-Drive	26	20	8.5
Ford Escape	4WD Hybrid	27	30	6.5
Saturn Vue	6-cylinder auto	23	16	9.8
Saturn Vue	Hybrid	32	25	6.7

Note: The US EPA and the US Department of Energy calculated these figures based on an average of 55% city driving and 45% highway driving for 15,000 miles per year.[64]

Switching from a gas-combustion model to a comparably-sized, cleaner hybrid vehicle is among the most familiar recommendations for consumers who want to reduce their carbon footprints. However, the figures in Table 1 show that the changes in annual tons of GHGs discharged per vehicle are not very substantial for most comparable vehicle models. The shift from a Camry 4-cylinder model to a Camry

Hybrid would retain 76% of the original GHG emissions; the shift to a Ford Fusion Hybrid would retain 49% of the gas-consumption model's GHG pollution; the shift to the Escape Hybrid would retain 76% of the GHG discharges; the shift to the Saturn Vue Hybrid would retain 68% of the previous GHG emissions; and the shift to a Prius Hybrid would retain about 53% of GHG discharges when compared to the similar mid-sized Camry 4-cylinder model. In effect, the shift from these gas-combustion models to hybrid vehicles for comparable models would leave appreciably more than half of the GHG pollution discharged by non-hybrid BAU models. These residual GHGs will combine with the existing "stock" of GHG pollution in the air to increase, not reduce, the cumulative atmospheric GHG concentration and related greenhouse effect.

Switching to hybrid models would undoubtedly *reduce* annual GHG emissions from new vehicles in comparison to absolutely no BAU reductions, but the BAU baseline should not be used. The residual discharges from the hybrid models will *increase* the cumulative atmospheric GHG concentration and associated climate change risks. The only positive thing one could say about the conversion to the hybrid models shown in Table 1 is that this emissions-reduction measure would be a little better than nothing at all because it would delay the full impact of climate change harms from these vehicle emissions for a short while. For example, it would take roughly two years driving a Ford Fusion Hybrid to generate as much aggregate GHG discharges as one year driving a non-hybrid Fusion model. But this vehicle transition would still entail substantial residual discharges increasing the cumulative atmospheric GHG concentration, and therefore could not produce any significant climate change benefits. Rather, the residual GHG discharges after the shift to hybrid models would fail to reduce the atmospheric GHG level and inevitably will waste a very large amount of money, time, and efforts that could be expended on more effective motor vehicle mitigation programs.

In contrast to hybrids, many car manufacturers have announced their intention to market a variety of all-electric vehicles,[65] such as the Nissan Leaf and Chevy Volt, which are either ready now for commercial sales or are projected for widespread sales in the near future. Aside from the production-phase energy costs and pollution, which normally would apply roughly equally to fossil fuel-burning cars and to new GHG-free models, the clean GHG-free vehicles will not discharge residual GHG emissions and consequently they would be much more desirable than hybrids from a climate change perspective.[66]

Unfortunately, GHG-free vehicles will have an uphill fight against established fossil fuel combustion vehicles and hybrids because of the high production and operating costs during the early years, which are not being adequately subsidized by government funding as a collective pollution-control action benefiting most people no matter what vehicles they personally drive. The comparison between hybrids and clean electric vehicles is a good example of the desirability of focusing our efforts on GHG-free replacement technologies that do not generate any residual GHG emissions that will worsen the atmospheric greenhouse effect.

The US federal government invested tens of billions of dollars in building the Interstate Highway System half-a-century ago for less urgent reasons; but now, the government is not close to making comparable investments to promote GHG-free vehicles or the infrastructure needed to operate them. Most governments have not begun to establish electric-car recharging stations or battery-replacement stations.[67] Most governments are not using their purchasing power to create economies of scale and competition in the production of clean vehicles. And most governments are not adopting emissions-reduction regulations or carbon taxes that could create incentives for more consumers to buy GHG-free clean cars, which would help reduce the cumulative concentration of GHGs in the atmosphere for the collective benefit of everyone. As long as the purchase price and operating costs of fossil fuel-burning vehicles fail to incorporate the full array of externalized damages they are causing, including substantial contributions to climate change harms, it will be difficult for GHG-free vehicles to capture a large proportion of the motor vehicle market as a result of economic constraints more than technological limitations.

Persistent GHG Discharges

The damage from discharging large amounts of residual GHG emissions into the air each year is compounded by the persistence of the most common greenhouse gas, carbon dioxide. In an assessment of the scientific knowledge on global warming, the US Environmental Protection Agency (EPA) concluded: "The major greenhouse gases emitted by human activities remain in the atmosphere for periods ranging from decades to centuries. It is therefore virtually certain that atmospheric concentrations of greenhouse gases will continue to rise over the next few decades."[68] The UN Intergovernmental Panel on Climate Change (IPCC), the largest and most influential international association of scientists who study climate change problems, similarly concluded: "The consequences of greenhouse gas emissions changes, and CO_2 emissions changes in particular, take decades to centuries to manifest themselves fully in the climate system."[69] Given this high degree of persistence, GHGs discharged into the atmosphere in recent years are already in the global warming "pipeline" and may continue to increase global warming and climate change for centuries to come.[70]

A little more than a decade ago, several scientists from the American Geophysical Union concluded: "it is now generally believed that a substantial fraction of the excess CO_2 in the atmosphere will remain in the atmosphere for decades to centuries, and about 15–30% will remain for thousands of years."[71] More recent research found that the extent of GHG persistence has been underestimated in prior studies, and the atmospheric warming effects of carbon dioxide may often last for thousands of years rather than centuries.[72] It is also true that "reductions of other greenhouse gases cannot compensate for the long-term effects of emitting CO_2."[73]

The persistence of CO_2 discharged today and in the next several decades will adversely affect the atmospheric GHG concentration centuries from now, and will contribute to diverse climate-related dangers that billions of people will experience in the near and distant future.[74] A clear understanding of GHG persistence is essential for evaluating the climate-policy ramifications of residual GHG pollution discharges or cap-and-trade allowances. The need to understand the implications of CO_2 persistence is especially vital because nearly all consensus mitigation programs have adopted back-loaded emissions-reduction targets that will not seriously begin to curtail GHG pollution levels for several decades. Consider the effects of persistence under the Waxman–Markey Bill and post-Copenhagen Obama Administration promises:

Table 2: Waxman–Markey/Obama GHG Emissions-Reduction Targets

Target Dates	Mandated % Cuts	Residual % Allowed	Annual US Discharges
Up to 2020	0%	100%	~6.00 billion tons
2020–2025	17%	83%	~4.98 billion tons
2025–2030	30%	70%	~4.20 billion tons
2030–2050	42%	58%	~3.48 billion tons
After 2050	83%	17%	~1.02 billion tons

Assuming that carbon discharges from the US were about 6 billion tons in 2005,[75] which may somewhat understate the total amount of annual GHG emissions,[76] the last column in Table 2 shows the billions of tons of annual residual GHG pollution authorized by the various stages of these emissions-reduction programs. This table does not take into account the likely possibility that US GHG discharges will expand during the designated time periods in response to increasing population growth, economic growth, or changing consumption patterns. The residual pollution figures may be somewhat exaggerated because some polluters will reduce their GHG emissions before they reach the designated target dates; but on the other hand our previous pollution control experiences suggest that there will be many laggard firms and widespread non-compliance. The figures in Table 2 show that the volume of persistent residual GHG discharges that would have been authorized under these legislative and administrative initiatives would greatly outweigh the mandated emissions reductions until after the final target date of 2050.

The Waxman–Markey Bill and the Obama Administration post-Copenhagen promises are based on adopting cap-and-trade programs under which the persistent residual discharges would be authorized as tradable GHG allowances or carbon permits. The GHG allocation formula or mechanism for the cap-and-trade allowances would have major distributional effects depending on whether these allowances would be acquired primarily by auction sales or by free "grandfathered" give-aways to polluting industries, as the Waxman–Markey Bill provided. Yet, from a global warming perspective, the distributional effects are relatively unimportant because the emissions associated with permitted GHG allowances will be discharged into the atmosphere in exactly the same manner as persistent residual GHGs and will exacerbate the atmospheric GHG concentration largely to the same degree, no matter which companies or agencies control the cap-and-trade allowances.

In essence, persistent residual GHGs are the GHG emissions discharged into the atmosphere after the applicable emissions-reduction targets are met, and GHG cap-and-trade allowances would be discharged into the atmosphere in precisely the same manner after the regulatory cap is met under a cap-and-trade system. In both emissions-reduction contexts, the "permissible" residual GHG pollution authorized by regulation or by cap-and-trade allowances would be discharged into the atmosphere, thereby increasing the cumulative GHG concentration in exactly the same way.

This back-loaded mitigation pattern is a serious climate-policy mistake when evaluated in terms of the persistence of CO_2 affecting the atmospheric GHG concentration for centuries or longer. For every ton of BAU pollution cutbacks achieved by the consensus emissions-reduction programs, several or many tons of persistent residual GHG emissions coming from GHG direct regulation or cap-and-trade allowances will be discharged into the atmosphere and will worsen climate change risks for a very long time. Even the most stringent emissions-reduction target of 83% after 2050, which is more fantasy than regulation, would allow 17% of the 2005 GHG pollution volume to become persistent residual discharges that will increase the already-existing atmospheric GHG concentration while compounding the harmful greenhouse effect and climate change dangers. Emissions-reduction programs that will not significantly *reduce* BAU discharges until decades into the future while allowing persistent residual emissions or discharges authorized by tradable allowances and carbon offsets to *increase* the cumulative atmospheric GHG concentrations this year and every succeeding year are certain to cause severe long-term climate degradation.

The "reducing the increases" and CO_2 persistence problems are equally applicable on an international plane. The Kyoto Protocol, for example, provides that the participating developed nations (Annex I states) should cut their GHG emissions to roughly 5% below 1990 discharge levels on average, and only the United States among affluent countries has not agreed to meet this minimal emissions-reduction target. However, even if all the nations that ratified the Kyoto Protocol meet their self-assumed commitments by 2012, the treaty member-states will be allowed to continue discharging all but approximately 5% on average of the very large BAU quantities of GHG pollution these nations were putting out in 1990.[77]

It would be hard to find a better example (or worse one) of a "reducing the increases" program that obligates participating developed nations to *reduce* GHG discharges by a minimal amount while allowing them to continue discharging great volumes of persistent residual GHG emissions that are bound to *increase* the cumulative atmospheric GHG concentration. Because virtually all policymakers have been using the BAU baseline rather than the atmospheric concentration baseline, they may consider the very small emissions reductions achieved by the Kyoto Protocol as a worthwhile though insufficient achievement.

On the other hand, if the atmospheric GHG concentration baseline is adopted, as it should be, the Kyoto Protocol international pollution control program should be regarded as a disastrous climate-policy mistake that has allowed the rapid growth of persistent residual GHG discharges steadily increasing the aggregate atmospheric GHG concentration and related climate change dangers. The Kyoto Protocol should be regarded as a terrible climate-policy mistake because it creates the illusion of emissions-reduction progress while in reality it has allowed a devastating expansion of the atmospheric GHG concentration and the greenhouse effect. This ineffectual international agreement allows participating developed countries to appear satisfied with meeting very minimal emissions-reduction targets that have had no beneficial effects on the growth of climate change risks in the past decade.

In a similar vein, the European Union "20–20–20" plan indicates they intend to reduce their GHG emissions 20% by 2020, but this back-loaded emissions-reduction policy would allow the remaining 80% of GHG emissions in 2020 and for decades thereafter to become residual discharges that will adversely affect the atmospheric concentration and the greenhouse effect.

Even if the EU decides to implement a 30% reduction rate that they have been discussing in recent years,[78] this supposedly radical plan would still eliminate less than a third of the annual GHG pollution the European countries would be discharging. The remainder of the GHGs would be pumped into the air as persistent residual discharges or as GHG allowances under their Emissions Trading Scheme cap-and-trade system. The vast quantities of GHGs discharged under either European emissions-reduction regulations or ETS allowances would clearly increase, not stabilize or decrease, the atmospheric GHG concentration that is causing the greenhouse effect and climate change.

What climate change benefits will we gain from consensus emissions-reduction programs in the next 10 years, 30 years, 50 years, or 100 years? It is important to emphasize that the 95% of residual GHG emissions from the Kyoto Protocol, the 90% of residual GHG emissions from the RGGI program, the 85% of residual emissions from the Western Climate Initiative (WCI), the 70% of residual emissions from the California and federal "clean car" regulations and the currently-debated EU post-Kyoto regulations, the 50% of residual emissions from the 2008 G8 agreement (if this commitment is ever implemented, which I doubt), and the residual GHG discharges allowed by all other consensus emissions-reduction programs are *persistent* discharges that may remain in the atmosphere for many centuries or millennia. If the atmospheric GHG concentration continues rising as a result of the addition of more residual GHG discharges or a high number of cap-and-trade allowances authorized by the consensus emissions-reduction programs, no appreciable climate change benefits would be achieved despite many billions of dollars in GHG pollution-control expenditures and many possible consumption restrictions.

Moreover, the insufficient emissions-reduction targets in future decades are far from credible pollution-control commitments. In his recent book on climate change dangers, Dr. James Hansen, the NASA climate scientist who has been a pioneer in publicizing global climate change dangers and who courageously resisted the attempts of Bush Administration officials to muzzle his opinions, has contended that the longer-range emissions-reduction promises in the Waxman-Markey Bill and other legislative proposals are "*meaningless.*"[79] Hansen suggested asking the politicians who have sponsored these programs whether their plans will cut out future coal-combustion processes, will stop building new coal-burning power plants, and will find ways to gradually curtail society's reliance on oil or other fossil fuels. Without major transformations in reducing fossil fuel usage, there would be no way to achieve the ambitious GHG cutbacks promised by virtually all emissions-reduction proposals. Dr. Hansen contended that when the politicians claim they are going to overcome climate change by adopting "a goal, binding target, or a cap," without defining specific measures to limit fossil fuel exploitation: "you know they are lying. Yes, *lying* is a harsh word, so you may instead say 'kidding themselves'. But I expect that one day your more perceptive grandchildren will say that you let the politicians lie to you."[80]

This may appear to be an extreme condemnation aimed at some well-meaning legislators, including Representatives Waxman and Markey, and yet it is difficult to see what realistic basis these politicians have for promising massive GHG cuts several decades in the future while we continue to increase our fossil fuel combustion and harmful GHG discharges every year. By 2050, current and future GHG-polluting activities will have put out so much persistent residual discharges that only an extremely draconian emissions-reduction plan would have any chance of being more than a fairy tale promise of mitigation. I must agree with Dr. Hansen that what the American people have been receiving from their leaders is partly wishful thinking, partly ignorance, partly attempts to avoid or deny their policymaking responsibilities, and partly deliberate misrepresentations in the guise of "deal making" favoring special interests that will benefit economically from continuing to degrade climate conditions despite the severe long-term dangers to billions of people.

The only arguable benefit of "reducing the increases" emissions-reduction programs is that the greenhouse effect, global warming, and climate change might eventually become even worse if we do absolutely nothing to restrict GHG pollution. But a lesser "bad" does not make a "good" or a good climate policy. We cannot tolerate back-loaded emissions-reduction programs that will compound atmospheric GHG levels and worsen climate change for centuries to come, based on the foolish argument that conditions would be even more harmful if we do nothing at all. We must devise climate protection strategies that are much better than "nothing" and are also better than "reducing the increases" programs, which cannot come close to keeping us where we are now in terms of the atmospheric GHG concentration, to say nothing of their inability to limit future climate change dangers. Despite increasingly impassioned warnings from a great many scientists and professional environmentalists, the climate-policy mistakes the world's leaders have been making are bound to produce widespread mitigation failures with consistently disastrous consequences.

Both proponents and opponents of the consensus back-loaded emissions-reduction approach believe that these pollution-control programs could succeed in reducing climate change risks if they are properly implemented, but the opponents normally contend that the economic and social costs of these regulations would be too high. In contrast, I am arguing that the consensus emissions-reduction approach cannot achieve significant climate change progress or any tangible mitigation benefits as a result of their back-loaded design, massive residual GHG discharges, and the long-term persistence of carbon dioxide. Climate policymakers and other concerned people should recognize the distinction between a criticism that contends effective mitigation programs may or may not be worth the emissions-reduction costs, in contrast to my conclusion that we have been choosing foolish approaches that cannot possibly achieve effective mitigation outcomes no matter how large the fiscal investments and administrative resources expended on efforts to make the consensus emissions-reduction programs function properly. Climate policymakers and their advisors must learn to understand the "reducing the increases" emissions-reduction problem, the "BAU baseline" versus "atmospheric GHG concentration baseline" problem, and the "back-loading of persistent GHG discharges" problem before they can choose more thoughtful climate policies.

Natural and Human-Made "Sinks"

Some readers of previous drafts of this chapter have asked why, if we are able to reduce business-as-usual GHG pollution by a large percentage, perhaps 80% or more, the remaining residual GHG emissions will not be eliminated or controlled by natural absorption and sequestration processes. However, we must again consider the "stocks and flows" cumulative GHG concentration problem. The atmospheric GHG concentration baseline (level **C** in Figure 1) represents the current GHG concentration in the air, which is *already too high*[81] and has been causing numerous harmful climate change impacts, including melting glaciers and ice sheets, droughts, torrential rains and floods, stronger storms, more wildfires and tornados, etc. These climate-related dangers are occurring *now* at the present atmospheric GHG concentration, which means that the natural processes of greenhouse gas absorption or sequestration by the oceans, forests, and agricultural "sinks" are insufficient to overcome the climate change impacts arising from the existing atmospheric GHG concentration.

If we adopt back-loaded "reducing the increases" emissions-reduction programs that will allow large amounts of persistent residual discharges or comparable amounts of GHG discharges from cap-and-trade allowances to go on increasing the atmospheric GHG concentration (level **C** plus level **R** in Figure 1), it is difficult to imagine how the already-inadequate natural sinks[82] could absorb enough GHG discharges in the future to overcome additional persistent heat-trapping GHG pollution.[83]

An equally great cause for concern is that the capacity of natural sinks to absorb more GHGs appears to be shrinking rather than expanding. The warming of surface ocean waters resulting from higher atmospheric temperatures is reducing the volume of GHGs that the ocean is able to absorb, leaving more GHGs in the air.[84] The increasing acidification of ocean waters may also reduce the amount of additional CO_2 that the seawater and marine life can absorb.[85] Many human activities are degrading sea grasses, mangrove forests, and other forms of marine plant life that may serve as carbon sinks.[86] Deforestation remains prevalent in developed and developing countries,[87] while agricultural lands are regularly lost to worldwide urbanization trends. In light of these changes, the currently insufficient natural sinks cannot be expected to keep up with the ongoing growth of the atmospheric GHG concentration. If anything, natural sinks are becoming less effective while the need for them becomes greater.[88]

For millions of years there have been various natural sources of CO_2 discharges, but they were small enough for natural sinks to absorb them — this process is often called the "carbon cycle." However, the carbon cycle has been broken by countless human activities. The atmospheric greenhouse effect has been substantially worsened by increasing human discharges of heat-trapping gases, especially CO_2, in the past century that have been much larger than prior natural GHG pollution levels and that are clearly too large for the natural sinks to absorb.

Fortunately, we do not need to reach zero CO_2 emissions, which we cannot possibly do, to overcome global climate change risks. Imposing an impossible emissions-reduction target would merely be cause for greater despair. Instead, we must decrease aggregate CO_2 emissions until they are lower than the cumulative absorption capacity of natural sinks and potential human-made sinks. This goal will be difficult and expensive to achieve, but it is neither impossible nor unaffordable when we consider the likely climate change dangers that could be avoided. The world's leaders must recognize that nature is not going to rescue us from worsening climate problems, and therefore we will have to rescue ourselves or suffer the consequences. A workable climate change solution has been precluded more by selfish, short-sighted, uninformed, sometimes corrupt political leadership, by callous special-interest lobbying, and by collective-action problems including free-rider resistance than by technological or natural limitations.

It is possible that we could develop human-made carbon sinks enabling the generation of more GHGs without more climate damage.

Many business and agency research programs have been exploring the feasibility of capturing GHG emissions from fossil fuel combustion and pumping the harmful gases deep into the earth or ocean to prevent them from reaching the atmosphere.[89] This Carbon Capture and Storage (CCS) process may prove extremely important for combating climate change in future decades because, if it could become sufficiently reliable, affordable, and environmentally safe, it would enable many nations to continue relying on fossil fuels for energy while reducing further damage to the climate.[90]

This is a set of very big "*ifs*" and yet, if all the "*ifs*" can be overcome, we could potentially increase the world's GHG production substantially without additional climate harms as long as fossil fuel combustion gases do not reach the atmosphere and further degrade climate conditions. This CCS illustration shows that tracking annual BAU pollution cannot determine whether climate change is getting better or worse. Instead, we must look at the effects of GHG discharges on the atmospheric GHG concentration to make this critical finding.

Regrettably, in mid-July 2011 the American Electric Power Company (AEP) chose to cancel the construction of the first full-scale American CCS plant in West Virginia, where the company had operated a successful CCS pilot-project for the past two years.[91] The company blamed the termination of the project on economic constraints, but government officials and other experts contended that AEP was no longer under any pressure to reduce its GHG discharges after the Republicans won control of the House of Representatives in 2010.[92] In one report, "A senior Obama administration official said that the AEP decision was a direct result of the political stalemate."[93] Little climate change mitigation progress can be expected on any front as long as the US government is paralyzed by factional in-fighting, and sadly there is no reason to believe that the current political conflicts will be resolved in the near future. The failure of this full-scale CCS project is another example supporting my contention that the leading difficulties in responding effectively to climate change mitigation problems are political and economic, not technological and ecological.

Reducing the Growth Rate of GHGs in the Atmosphere

I prefer the phrase "reducing the increases in the atmospheric GHG concentration" to "reducing the growth of the atmospheric GHG concentration." In practice, however, they amount to exactly the same thing. The word "growth" has many positive connotations, but there is nothing desirable about the growth of the atmospheric GHG concentration or the resulting growth of the greenhouse effect and climate change harms. Here are several analogies that may help explain the growth of climate change risks and the corresponding dangers from the climate-policy mistake of relying on "reducing the increases" emissions-reduction programs that cannot produce any tangible climate change benefits.

When President Bush in April 2008 announced a lame-duck goal of reducing the growth of American GHG emissions 18% by 2012 and completely stopping the emissions growth by 2025,[94] a columnist for the *New York Times*, Gail Collins, cleverly observed that if President Bush had adopted an equivalent "reducing the growth rate" approach to his body weight when he took office in 2000, he would have weighed 332 pounds in 2008, 400 pounds in 2012, and 486 pounds in 2025 when American GHG growth is supposed to cease under his plan.[95]

This satirical image of the "fat Bush"[96] becoming steadily more corpulent while "reducing" his obesity growth rate is essentially the same as the "reducing the increases" GHG emissions-reduction approach I am criticizing. Each additional pound does not represent a separate danger that can be curtailed, but rather the cumulative effect of continuing to add more and more weight, compounding the President's hypothetical obesity problem. The unsatisfactory practice of only reducing the GHG emissions growth rate, not the cumulative greenhouse gas volume in the atmosphere, applies not only to President Bush's waistline but to all of the consensus emissions-reduction programs mistakenly advocated by climate policymakers and environmental groups as ostensible "solutions" for climate change problems.

In *The Rough Guide to Climate Change*, Robert Henson noted: "Reducing greenhouse gas emissions by a few percent over time is akin to overspending your household budget by a decreasing amount each year; your debt still piles up, if only at a slower pace."[97] Again, decreasing the growth rate of the GHG pollution emissions is not sufficient to prevent increasing the aggregate GHG concentration underlying the greenhouse effect and climate change risks.

Andrew Revkin, a *New York Times* reporter, similarly observed: "As I and others have put it, slowing emissions of carbon dioxide is somewhat like slowing credit-card spending and expecting your debt to shrink."[98] The consensus emissions-reduction programs may cut the volume of new GHG discharges to some extent when compared to the absence of any BAU pollution-control program, but the residual emissions will be added to the large "stock" of GHGs that has already been absorbed by the atmosphere. Combining these new GHGs with the existing GHGs in the air will cause the greenhouse effect to grow worse every year.

In another column, Mr. Revkin quoted the "greenhouse bathtub effect" explanation of Professor Sterman: "Basically, the atmosphere is like a bathtub with a partially opened drain. Carbon dioxide from burning fuels and forests is flowing in twice as fast as it is being absorbed by plants and the ocean, and some of those 'sinks' are in fact getting saturated, it appears, meaning that the 'drain' is clogging a bit."[99] Professor Sterman warned: "Because the drains out of the various bathtubs involved in the climate — atmospheric concentrations, the heat balance of the surface and oceans, ice sheet accumulations, and thermal expansion of the oceans — are small and slow, the emissions we generate in the next few decades will lead to changes that, on any time scale we can contemplate, are irreversible."[100] This is precisely what the recent congressional bills and Obama Administration back-loaded emissions-reduction proposals would allow: enormous persistent GHG discharges in the next few decades that will cause irreversible damage to the climate and the countless people who depend on benign climate conditions.

Even if the flow from the "bathtub faucet" is appreciably decreased, as long as the flow into the bathtub remains greater than the drain's capacity, the water level will continue rising until it overflows the bathtub and creates a flood. *National Geographic* magazine recently published several pages of graphics illustrating the "bathtub effect" with explanatory text, and the article's conclusion was: "As long as we pour CO_2 into the atmosphere faster than nature drains it out, the planet warms. And that extra carbon takes a long time to drain out of the tub."[101]

The consensus emissions-reduction programs I am challenging would cut BAU pollution levels to some degree in comparison with no emissions restrictions whatever, but these pollution-control programs will allow greater volumes of persistent residual GHGs to be discharged into the atmosphere than can be eliminated through the "drains" provided by natural sinks.[102] As a result, these back-loaded emissions-reduction programs cannot improve the atmospheric greenhouse effect or decrease climate change hazards in any effective manner. Instead, the consensus national and international mitigation programs will continue to waste many billions of dollars by imposing "too little, too late" emissions-reduction plans that will not be able to prevent or significantly slow down the expansion of climate change hazards.

Unfortunately, this deferred-reduction or back-loaded approach is exactly what most concerned American politicians have been proposing, with the preferred alternative policy option emerging as total climate neglect. Despite many impassioned warnings from scientists and environmentalists, our national and world leaders have been making climate-policy mistakes that are sure to produce numerous mitigation failures leading to societal and environmental disasters.

Persistent residual GHG discharges are the underappreciated arch-villains of climate change regulation. Few, if any, concerned policymakers and expert advisors have considered this problem when they advocate back-loaded emissions-reduction programs. They are looking at the BAU pollution baseline, and they all want to cut annual BAU discharges by a selected percentage rate — making the serious climate-policy mistake of imagining that cutting annual GHG discharges will result in comparably reducing climate change risks. These people are not even looking at the cumulative atmospheric greenhouse effects of persistent residual GHG emissions that would be authorized by their proposed climate regulations or by cap-and-trade allowances and carbon offset programs. By ignoring the long-term persistence and cumulative effects of residual CO_2 discharges, the consensus climate policy is creating a "Sword of Damocles" that will hang over our descendants for hundreds or thousands of years.

I have often wondered why scientists working on climate issues have not recognized the problems discussed here or acknowledged that the consensus emissions-reduction proposals they have been supporting will yield no significant climate change benefits. The Union of Concerned Scientists (UCS), for example, has contended that: "If we assume the world's developing nations pursue the most aggressive reductions that can reasonably be expected of them, the world's industrialized nations will have to reduce their emissions an average of 70 to 80% below 2000 levels by 2050."[103] After receiving a barrage of solicitation letters from UCS asking me to support their work on climate change, I sent an early draft of this chapter to them and asked why they are following the consensus emissions-reduction strategy[104] when this approach cannot succeed in producing any notable climate change progress.

I received a reply from Dr. Brenda Ekwurzel, a UCS climate scientist, saying that they are aware of persistence problems and citing passages in their publications that describe these concerns. She continued: "*Hence the UCS policy goals are aimed to limit the peak atmospheric concentration that the world will endure*.... Thanks for delving deeper into this topic and for helping to share your knowledge with others that we may not be currently reaching."[105]

I do not agree that UCS has been publicizing anything resembling the message in this book: UCS does not mention the negative effects of residual GHG discharges from the consensus emissions-reduction proposals, including their own proposal; the unsatisfactory results of "reducing the increases" emissions-reduction targets, such as the ones they are advocating; the inability of cap-and-trade systems to overcome "reducing the increases" failures; or the absence of any concrete climate change benefits from the emissions-reduction program UCS advocates despite the recommended expenditure of billions of dollars for completely ineffectual results. Instead, UCS continues to tell their members and contributors that meeting their proposed emissions-reduction targets will substantially help to overcome climate change risks, which is untrue and the direct opposite of the position argued in this book.

Dr. Ekwurzel identified the major projected benefit from the UCS "reducing the increases" proposal as limiting "the peak concentration" of GHGs in the air. This appears to be no more than an attempt to reduce the GHG concentration below what it would be if we do nothing at all to curtail GHG discharges. In other words, the UCS proposal entails only a "better than nothing" rationale. Their proposed approach would still allow 20–30% of persistent residual GHG discharges from developed nations to reach the air annually after 2050 and a much larger percentage of persistent residual GHG discharges before 2050. This means the atmospheric GHG concentration would continue getting steadily worse rather than "peaking."

From a dynamic perspective, the UCS recommendation could not actually limit "the peak concentration" of GHGs in the atmosphere; all it would do at best is to increase by a little while the interval of time required for the GHG concentration to reach a higher level. If the emissions-reduction targets supported by UCS were adopted and somehow met all of their goals, annual GHG pollution would be cut roughly two-thirds by 2050 and the same volume of persistent residual GHGs would be discharged into the atmosphere every three years as would be discharged in one year if we do nothing to restrain BAU pollution. If the atmospheric GHG concentration continues increasing, even at a slower rate than under a BAU scenario, the UCS proposal and other consensus emissions-reduction programs will not stabilize the atmospheric GHG level at any "peak concentration" and will never overcome climate change risks.

The most that the UCS emissions-reduction proposal could achieve, if it were adopted and faithfully implemented, would be to slow down the continuing growth of the atmospheric GHG concentration by a relatively modest interval of time. As with other consensus emissions-reduction programs, the UCS proposal would allow the GHG concentration in the air to keep growing over time, which means that the greenhouse effect would keep getting commensurately worse, even when it takes a couple more years to generate the predicted climate change disasters than if we do nothing at all. Extending the time to the predicted climate catastrophes by a short interval, while achieving no actual mitigation benefits, is not a climate policy that UCS should consider acceptable in any way, much less advocating a consensus emissions-reduction approach that at best would have precisely this unsatisfactory effect.

It is not easy to modify the way people think about a problem once they have internalized a mistaken conception widely shared by others in their field. Climate policymakers need to stop believing that any reduction in business-as-usual GHG emissions will correspondingly reduce climate change risks. They need to stop thinking that if many people reduce their carbon footprints by some extent, those decentralized GHG emissions cuts would produce significant progress toward mitigating climate change dangers. They need to stop thinking that reducing annual GHG pollution discharges will eliminate a commensurate proportion of climate change risks, rather than only extending the timeframe by a few months or years before comparably severe climate-related harms materialize. These leaders also need to stop believing that "every little bit helps," and instead they must recognize that more ambitious mitigation efforts with high worldwide costs will be required to achieve any meaningful climate change progress.

While persistent residual GHG discharges are increasing the cumulative atmospheric GHG concentration, how could the feeble back-loaded emissions-reduction programs already adopted, such as the Kyoto Protocol, or the recently proposed US legislative initiatives and environmental group recommendations achieve meaningful progress in overcoming climate change perils? How can we resolve climate change problems by allowing them to become worse every year? How can the political sponsors of costly but ineffectual mitigation programs retain widespread public support when they will not be able to show any tangible climate change benefits from their emissions-reduction regulations or cap-and-trade systems? I wish many more climate policymakers and advisors were asking themselves these basic and yet vital questions.

The Two Degrees Celsius Non-Solution

After the publication of the IPCC's Fourth Assessment Report (AR4) in 2007, many nations shifted their GHG emissions-reduction focus from a percentage-rate cutback model to a proposal for restraining the growth of the global mean temperature to no more than two degrees Celsius (2°C) in comparison to pre-industrial temperature levels.[106] At the 2009 G8 meeting in L'Aquila, Italy, the leaders of the major developed countries accepted the 2°C target and also agreed to cut their GHG emissions 80% by 2050 in order to meet the new target.[107] The G8 leaders said there was a "broad scientific view" that the global mean temperature "ought not to exceed 2 degrees Celsius" above pre-industrial levels,[108] which was consistent with the widely-held but mistaken view that climate scientists have found the 2°C target is reasonably safe for people and the environment.[109] This politicized view of a 2°C GHG impact limit was reinforced by the Copenhagen Accord negotiated at the end of 2009:

> To achieve the ultimate objective of the [UNFCCC] Convention to stabilize greenhouse gas concentration in the atmosphere at a level that would prevent dangerous anthropogenic interference with the climate system, we shall, recognizing the scientific view that the increase in global temperature should be below 2 degrees Celsius, on the basis of equity and in the context of sustainable development, enhance our long-term cooperative action to combat climate change....
>
> We agree that deep cuts in global emissions are required according to science, and as documented by the IPCC Fourth Assessment Report with a view to reduce global emissions so as to hold the increase in global temperature below 2 degrees Celsius, and take action to meet this objective consistent with science and on the basis of equity. We should cooperate in achieving the peaking of global and national emissions as soon as possible, recognizing that the time frame for peaking will be longer in developing countries and bearing in mind that social and economic development and poverty eradication are the first and overriding

priorities of developing countries and that a low-emission development strategy is indispensable to sustainable development.[110]

The first problem with the adoption of a 2°C mitigation target is that an increase of 2°C compared to pre-industrial temperatures is far from reasonably safe and instead is likely to produce many climate change disasters. I have not found any statements in the AR4 in which IPCC scientists contended that a 2°C increase in the Earth's average temperature would be *safe*. Rather, Chapter 19 observed: "Global mean temperature changes of up to 2°C above 1990–2000 levels would exacerbate current key impacts, such as those listed above [loss of glaciers, increases in human mortality, and increases in the frequency and/or intensity of extreme events] (high confidence), and trigger others, such as reduced food security in many low-latitude nations (medium confidence)."[111] Chapter 19 predicted: "A warming of 2°C above 1990 levels will result in mass mortality of coral reefs globally... with one-sixth of the Earth's ecosystems being transformed... and about one-quarter of known species being committed to extinction."[112] This AR4 chapter also predicted: "There is high confidence that warming of 1 to 2°C above 1990–2000 levels would include key negative impacts in some regions of the world (e.g., Arctic nations, small islands), and pose new and significant threats to certain highly vulnerable population groups in other regions (e.g., high-altitude communities, coastal-zone communities with significant poverty levels)."[113]

In a similar manner, Chapter 20 of the AR4 predicted that: "most species, ecosystems and landscapes would be impacted by increases of global temperature between 1 and 2°C above 2000 levels"; and "between 550 and 900 million people [will suffer] an increase in water-related stress. In this case, the range between estimates represents the effect of different changes in rainfall patterns for a 2°C warming."[114] The cited passages in these AR4 chapters refer to somewhat different baseline years, but that should not dilute their conclusions that a 2°C increase in global temperatures cannot be characterized as "safe" for countless people and ecosystems.

The global median temperature is now estimated to be about 0.6 to 0.8 degrees Celsius above the pre-industrial level, which means that the 2°C target would involve approximately a 1.5 times greater increase in the average temperature differential during the next few decades compared to the total increase during the past two centuries. Consider that in the past couple of years we have already had a large number of devastating weather-related experiences, including the Pakistan and Australian floods, Russian heatwave, South China and East Africa droughts, California and Arizona wildfires, Myanmar cyclones, and similar extreme weather events.[115] In light of the hazardous climate conditions at the current worldwide median temperature, it is hard to imagine how a relatively larger expected temperature increase of about 1.2–1.4 degrees Celsius above the present average could be regarded as "safe." It is also important to add that limiting the median global temperature increase to 2°C does not exclude much more dangerous temperature deviations in particularly vulnerable locations.[116]

The AR4 Summary for Policymakers presented a graph showing six alternative climate stabilization points reflecting different global mean temperature ranges corresponding with six ranges of potential atmospheric GHG concentrations. These predictions offered a "best estimate" climate sensitivity of 3°C, an "upper bound" of likely climate sensitivity of 4.5°C, and a "lower bound" of likely climate sensitivity of 2°C.[117] The estimated atmospheric GHG concentrations for the first three stabilization scenarios were about 445–490 parts per million (ppm), 490–535 ppm, and 535–590 ppm, corresponding to average global temperature increases of 2°C, 3°C, and 4°C, respectively. The other three IPCC scenarios described even higher, more dangerous GHG concentrations and related temperature increases. This range of comparisons did not show that 2°C is "safe," but only that it is the lowest plausible temperature increase — not the most likely one — given the expected growth in the atmospheric GHG concentration.

Some of the scientists and policymakers who believe a 2°C mean temperature increase limit is attainable have argued that meeting this goal will require using an "overshoot" strategy. This would allow a temporary increase of the atmospheric GHG concentration up to around 550 ppm during the next couple of decades followed by a rapid decrease in global GHG emissions leading to the eventual restoration of the 450 ppm concentration corresponding to a 2°C mean temperature increase over pre-industrial levels.[118] This is really dangerous thinking, or perhaps wishful thinking, that would invite additional irresponsibility combined with additional climate disasters.

One scientific assessment claimed that: "A temporary, limited overshoot of greenhouse gas concentrations has only limited environmental implications. However, a more sustained and larger overshoot could lead to a more irreversible response."[119] A sharper critique of the risks presented by an overshoot strategy maintained that: "Faith in our ability to overshoot then return to a safer climate simply fails to understand the science — whatever we do we will be stuck with the results for a very long time. If carbon dioxide concentrations reach 550 ppm, after which emissions fall to zero, the global temperature would continue to rise for at least another century."[120]

The overshoot strategy depends on a level of international GHG emissions-reduction cooperation that does not yet exist and appears unlikely to exist anytime in the foreseeable future. As a consequence, I view this approach as another climate-policy mistake aimed at deferring controversial choices decades into the future while shifting the core mitigation and adaptation responsibilities to future generations of increasingly endangered people.

Another major problem with the 2°C mean temperature increase target, to say nothing of the overshoot strategy, is that it will be virtually impossible to achieve under current political conditions. The Copenhagen Accord, which invited all nations to make their own voluntary emissions-reduction commitments, was drafted and submitted to the UNFCCC member-states in December 2009. Since then, several studies have concluded that the national emissions-reduction commitments offered by the major GHG-polluting countries will not be stringent enough to come close to meeting the Copenhagen 2°C goal.[121]

The Secretary General of the Organization for Economic Co-operation and Development (OECD) recently noted that: "the most ambitious of the declared targets by industrialised countries add up to an 18% reduction in their emissions by 2020 (from 1990 levels).

This falls short of the 25 to 40% reduction that scientists say is needed to keep the temperature rise to 2°C above pre-industrial measures. We need to increase these targets."[122]

These speculative percentage reductions would require rapid cutbacks leading to a global peak in GHG emissions before 2020 and more rapid emissions reductions thereafter,[123] and yet the level of worldwide GHG pollution is not decreasing fast enough to approach these targets. There is not the slightest indication that worldwide GHG emissions are on course to peak within the next decade, which means the 2°C limit is just another false commitment in a long line of implausible wishful-thinking promises that will never achieve real climate change progress in the near or distant future.

The leaders of all major GHG-polluting nations are surely aware of the need for sharp cutbacks in their nation's GHG discharges if they are to maintain the 2°C limit, but these emissions reductions are not happening with anything like the necessary stringency. In some large GHG-polluting nations, including the United States and China, there appears to be no compelling political pressures that could ensure these states will undertake necessary emissions-reduction efforts within a timeframe consistent with limiting the increase in global mean temperature to 2°C. In other words, despite the supposedly "new" 2°C target, the political impediments in developed nations and developing countries are largely the same as with the previous mitigation efforts; and primary reliance on emissions-reduction commitments that some large GHG-polluting nations have refused to accept or implement is entirely the same as in the previous mitigation failures.

As illustrated by the quotation from the Copenhagen Accord, the developing nations put a higher priority on economic and social development than on climate mitigation efforts despite their vulnerability to climate change harms. This conflict is examined in detail in Chapter IV. It is worth noting, however, that the "25 to 40%" emissions-reduction target cited by the OECD Secretary General would require developing states to reduce global pollution substantially if the developed nations are only willing to meet the least stringent target of 25% GHG emissions reductions by 2020.[124] In reality, the presumed cutback by developing countries is completely at odds with the position they have adopted in international negotiations on climate change issues. Unless the developed nations are willing to make nearly the entire 40% reduction in worldwide GHG emissions reductions, there is little hope that the 2°C limit can actually be attained.

This shift in climate policy is little more than fantasy because the developed nations are nowhere close to meeting the lower-bound requirement of 25%, much less the higher 40% emissions-reduction target. Both of these target numbers are purely speculative guestimates based on questionable assumptions about the world's economic and political conditions during the next two decades. In essence, we do not know whether meeting these numerical targets would actually lead to climate change stabilization or progress, but we do know that neither the developed nations nor the major developing nations are seriously trying to achieve these target GHG reductions in practice.

When faced by complex climate change issues, people characterized as "contrarians" may nonetheless reach similar conclusions and proposed solutions. After contending that the 2°C limit cannot succeed, Oliver Geden, a German climate-policy analyst, argued persuasively that there would be little point in replacing the unrealistic 2°C target with another, less demanding, numerical emissions-reduction target that would be no more likely to succeed.[125] Instead, Dr. Geden argued for a "paradigm shift" replacing graduated GHG emissions-reduction programs with a global "carbon neutrality" strategy intended over the long-term to "reduce net emissions of greenhouse gases to zero."[126] He referred to this strategy as a "decarbonization" approach, which entails eliminating GHG sources whenever feasible.[127] Dr. Geden did not go into detail in his short paper aside from emphasizing the requirement that widespread decarbonization must be affordable. He did acknowledge that a decarbonization strategy would not guarantee the attainment of any specific global mean temperature in the future, but then he correctly argued that familiar emissions-reduction mitigation programs have not been succeeding in attaining reliable pollution control commitments and therefore they are also incapable of achieving any particular temperature range.

Dr. Geden did not mention the need to create clean replacement technologies or the technology transfer process of disseminating GHG-free technologies to developing countries to enable them to increase their economic welfare without damaging the climate, which are among the core factors in my proposals. Yet, his recommendation that we pursue the elimination of GHG sources to the greatest extent feasible, rather than picking yet another speculative emissions-reduction target, is compatible with my conclusions.

The IPCC Summary for Policymakers in the AR4 concluded: "There is high agreement and much evidence that all stabilisation levels assessed can be achieved by deployment of a portfolio of technologies that are either currently available or expected to be commercialised in coming decades, assuming appropriate and effective incentives are in place for their development, acquisition, deployment and diffusion and addressing related barriers."[128] This is an optimistic contention that we can move toward elimination or sharp reduction of GHGs if we convince the world's leaders that a decarbonization approach would be far more promising than the back-loaded emissions-reduction programs, which thus far have accomplished nothing except wasting a lot of money and time. Another way to look at the IPCC conclusion is these eminent scientists are contending that the necessary clean technologies are present or could be made present in the next few decades, and the major impediments to climate change solutions involve economic and political constraints, not technological ones.

To get anywhere, climate policymakers must focus on the negative effects of persistent residual GHG discharges, or persistent GHG cap-and-trade allowances and offsets, that would be authorized by all of the consensus GHG emissions-reduction programs. They also need to understand that the critical factor in climate-policy assessments must be the cumulative atmospheric GHG concentration. Viewed through this lens, the Waxman–Markey and Kerry–Lieberman Bills, the Obama Administration post-Copenhagen plans, the EU ETS cap-and-trade scheme, the projected revisions of the Kyoto Protocol, the positions of most major environmental groups, and all the other widely-embraced consensus GHG emissions-reduction programs would prove to be expensive failures that cannot achieve any significant climate change progress.

The most urgent need at this moment is to educate climate policy makers and their expert advisors about the self-defeating effects of "reducing the increases" programs in order to prevent them from committing the world's regulatory institutions and fiscal resources, as they have done in the past and are still doing, to wasteful climate-policy mistakes that cannot accomplish any of their professed goals. We cannot overcome climate change dangers by taking misguided steps down a costly mitigation road that is going nowhere.

the [illegible] of the many environmental groups, and [illegible] [illegible] [illegible] [illegible] [illegible] [illegible] [illegible] [illegible] [illegible] [illegible] [illegible] [illegible] [illegible] [illegible] [illegible] [illegible] [illegible] climate change [illegible]

The [illegible]

CHAPTER III

Economic Incentive Programs

Economic incentive programs, including cap-and-trade systems, carbon offsets, and GHG taxes, are widely supported in climate-policy circles on the grounds that they can achieve more efficient emissions-reduction outcomes. Firms with low pollution control costs will choose to curtail their emissions, while firms with high pollution control costs will purchase tradable allowances or offsets that enable them to continue their discharges at a cost, or they can choose to pay the pollution tax under a GHG tax scheme to avoid the imposition of more expensive technological emissions-reduction mandates. Under economic incentive mechanisms, GHG sources are able to decide for themselves based on their individual circumstances and profitability whether to reduce their discharges using current technologies, undertake research to develop innovative methods to decrease their regulatory costs, purchase tradable GHG allowances (or equivalent "tradable permits" or "pollution shares") to avoid expensive pollution control requirements, or pay the carbon taxes imposed under a GHG tax scheme.

Presumably, many thousands of major GHG sources will be able to make these decisions tailored to their particular circumstances better than government regulators could, and the GHG sources with low pollution control costs would be most likely to cut their emissions at the lowest feasible expense. The firms that develop better or cheaper pollution control methods can also profit by selling these technologies to other GHG producers with higher control costs or higher tax costs. The price of tradable allowances would serve as a market-signal of how valuable the right to continue discharging GHGs will be for various GHG sources and sectors.

This description of the theoretical advantages of economic incentive programs, including greater flexibility and the tendency to shift emissions-reduction costs toward a lowest-cost basis, underscores some powerful conceptual arguments in their favor. However, this overview and the large literature it summarizes does not address the central issue in this chapter — would the economic incentive programs in practice reduce the atmospheric GHG concentration in order to attain discernable climate change benefits?

Cap-and-Trade Systems

Widespread political and business support emerged in the past two decades for using cap-and-trade[129] systems to "put a market price" on GHG emissions and to reduce GHG pollution control costs in a supposedly more "efficient" manner than direct regulation can achieve.[130] Regrettably, the large literature debating the merits of cap-and-trade systems in comparison to technology-based direct regulation and pollution tax schemes has been inconclusive and often misleading. Advocates of reliance on cap-and-trade approaches focus almost entirely on the theoretical advantages and hypothetical efficiency gains from these economic incentive programs, while critics of cap-and-trade systems focus almost entirely on the difficult implementation, enforcement, and equity problems likely to arise. The proponents of cap-and-trade programs emphasize the theoretical economic benefits and asserted environmental benefits if everything goes right, but they rarely acknowledge the weaknesses that could undermine these programs if any implementation requirement goes seriously wrong.[131]

One indication that the positive cap-and-trade literature focuses more on cutting the costs of GHG emissions reduction than on increasing the climate benefits is that virtually the entire literature has been devoted to the operation of, and potential gains from, the trading process, while the vital environmental issue of how high or how low to set the regulatory cap has rarely been addressed.[132] Indeed, for many market-forces advocates, adopting a cap-and-trade system has become an end in itself rather than only a means for promoting lower-cost GHG pollution reduction. Thus, the large literature on cap-and-trade mechanisms reflects the familiar graphical image in Figure 2, in which the opposing arrows do not engage.

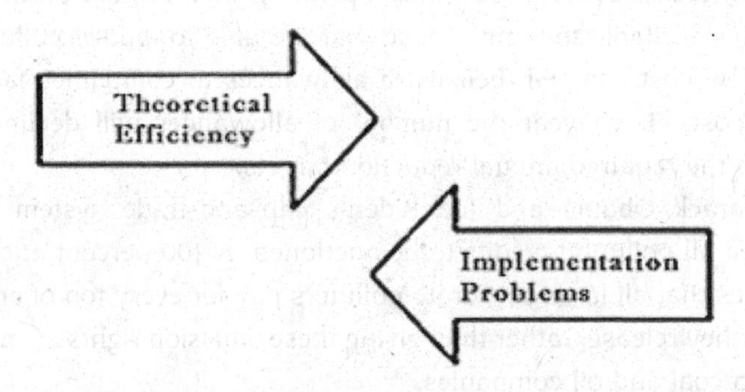

Figure 2: The Absence of Responsive Exchanges

Here, for example, is the justification presented by the Obama-for-President campaign for proposing an ambitious cap-and-trade program with a shrinking annual cap meant to reduce the volume of GHG discharges every year on the way to an 80% emissions-reduction target that would be imposed in 2050. This campaign statement was almost certainly written by expert advisors and then endorsed by the Democratic candidates responding to the idealized benefits, not the practical impediments, of a cap-and-trade strategy.

> Barack Obama and Joe Biden support implementation of a market-based cap-and-trade system to reduce carbon emissions by the amount scientists say is necessary: 80 percent below 1990 levels by 2050. They will start reducing emissions immediately in his administration by establishing strong annual reduction targets, and they will also implement a mandate of reducing emissions to 1990 levels by 2020.
>
> A cap-and-trade program draws on the power of the marketplace to reduce emissions in a cost-effective and flexible manner. Under the program, an overall national cap on carbon emissions is established. The emissions allowed under the cap are divided up into individual allowances that represent the permission to emit

> that amount. Because the emissions cap restricts the amount of pollution allowed, allowances that give a company the ability to pollute take on financial value. Companies are free to buy and sell allowances in order to continue operating in the most profitable manner available to them. Those that are able to reduce pollution at a low cost can sell their extra allowances to companies facing high costs. Each year the number of allowances will decline to match the required annual reduction targets....
>
> Barack Obama and Joe Biden's cap-and-trade system will require all pollution credits to be auctioned. A 100 percent auction ensures that all large corporate polluters pay for every ton of emissions they release, rather than giving these emission rights away for free to coal and oil companies.[133]

One year after the election, the Obama Administration accepted the Waxman–Markey Bill in place of their campaign proposal despite many weakening substantive changes. The revised program would use 2005 as the BAU-pollution baseline, authorizing a significantly larger volume of persistent residual GHGs than a 1990 baseline would. Most tradable allowances would be given away to grandfathered sources, not auctioned, and many would be given to "dirty" power plants, coal companies, and oil companies. Major corporate polluters would not "pay for every ton of emissions they release," although their customers who often lack any viable choices would probably have to pay for the GHG allowances. The emissions-reduction targets would not be decreased each year, but instead only during the four specified time intervals — 2020–2025, 2025–2030, 2030–2050, and after 2050 — as identified in Table 2 of Chapter II. And the GHG emissions-reduction target of 83% below 2005 discharge levels would be appreciably less stringent than a target of reducing GHG discharges by 80% below the 1990 GHG pollution level.

Despite the changes to many substantive details presented in the original campaign plan, the Obama Administration's acceptance of the cap-and-trade program and emissions-reduction targets in the Waxman–Markey Bill undoubtedly must have been based on the theoretical advantages of cap-and-trade, not on the actual or probable results, which are unknown. The US has had no experience with a cap-and-trade program regulating tens of thousands of GHG pollution sources, and also has had no experience with a very strict emissions-reduction cap, such as the 2050 target of 83% reductions from 2005 discharge levels. It is fair to say about any economic incentive program and any direct regulation emissions-reduction program that "the devil is in the details," but the Obama Administration does not know the specific details of how an ambitious cap-and-trade program for GHGs could or should be implemented. Rather, they know the *idealized theory* of cap-and-trade, and they have been emphasizing the hypothetical efficiencies far more than the practical implementation constraints.

At the other extreme, Dr. James Hansen asked President Obama to adopt a tax on fossil fuels rather than a cap-and-trade system because under the tax: "No lobbyists will be supported. Unlike cap-and-trade, no millionaires would be made at the expense of the public."[134] In the same manner, the noted investor and philanthropist, George Soros, predicted that under a cap-and-trade system "the trading will have all kinds of loopholes and misuse of the regulations and all kinds of ways of making money without actually dealing with the problem that it's designed to cure. So that's how the political process distorts things."[135]

The president of one energy company recently published an editorial in the *Washington Post*, saying: "If you liked what credit default swaps did to our economy, you're going to love cap-and-trade. Just read Title VIII [of the Waxman–Markey Bill], which lets investment banks, hedge funds and other speculators participate in the cap-and-trade market. They don't have emissions to cut; they have commissions to make."[136] And in virtually the same words, Thomas L. Friedman, the prominent *New York Times* columnist, predicted that: "cap-and-trade will be managed by Wall Street. If you liked credit-default swaps, you're going to love carbon-offset swaps."[137] These kinds of criticisms call into question how a cap-and-trade system for GHGs would actually function under realistic political and economic conditions, and who would really benefit from it — concerns far removed from theoretical discussions about how cap-and-trade programs could efficiently reduce the costs and rigidity of GHG pollution control choices.

During the congressional mark-up proceedings on the Waxman–Markey Bill, the critical predictions proved well-founded. One analysis of the legislative changes concluded: "The bill is far weaker than President Obama and the nation's environmental leaders had envisioned — So much so that Greenpeace, Friends of the Earth, Public Citizen and nine other environmental groups announced tonight that they cannot support the bill unless it is significantly strengthened."[138] This environmental protection bill was twisted so much by affected industries, special-interest groups, and their lobbyists and lawyers, that environmentalists became the most vocal opponents, though some large environmental groups continued to embrace the foolish idea that any weak climate change bill would be better than none. I strongly disagree with this counter-productive thinking in Chapter V. Among the criticisms voiced by the environmental groups that opposed the Waxman–Markey Bill were these complaints, which were entirely justified:

> The resulting bill reflects the triumph of politics over science, and the triumph of industry influence over the public interest.
>
> The targets are far less ambitious than what is achievable with already existing technology.
>
> They are further undermined by massive loopholes that could allow the most polluting industries to avoid real emission reductions until 2027. Rather than provide relief and support to consumers, the bill showers polluting industries with hundreds of billions of dollars in free allowances and direct subsidies that will slow renewable energy development and lock in a new generation of dirty coal-fired power plants.
>
> At the same time, the bill would remove the President's authority to address global warming pollution using laws already on the books.[139]

In 2010 and thereafter, the Obama Administration post-Copenhagen plan and the related congressional proposals have not been strengthened, and most commentators on climate change legislation believe that even the weak Waxman–Markey Bill and comparably weak Kerry–Lieberman Bill do not have sufficient political support in the Senate to overcome parochial cost-based concerns and special-interest opposition to cap-and-trade systems. Indeed, during 2010 the bills relying on cap-and-trade mechanisms were replaced by several other Senate bills that would not use a cap-and-trade approach at all or would limit it only to power plants and the fossil fuel energy sector.[140]

One legislative assessment concluded that cap-and-trade was "done in by the weak economy, the Wall Street meltdown, determined industry opposition and its own complexity."[141] This assessment also claimed that in an effort to gain political support, the Waxman–Markey Bill: "dished out a cornucopia of concessions and exemptions to coal companies, utilities, refiners, heavy industry and agribusinesses. The original simplicity was lost, replaced by a bazaar in which those with the most muscle got the best deals."[142]

Although cap-and-trade became wholly or mostly comatose in Congress during 2010 and 2011, it is still the primary mechanism in the European Union for achieving GHG emissions reductions. The conceptual efficiency of cap-and-trade systems may lead to renewed popularity for this economic incentive mechanism in the US once the present recession economy has significantly improved. I do not plan to offer a comprehensive assessment of the benefits and weaknesses of cap-and-trade systems, but rather to focus on whether a properly functioning GHG cap-and-trade system would actually improve climate change risks or would fail to achieve meaningful climate progress because it is subject to the same climate-policy mistakes discussed in Chapter II.

Under a cap-and-trade system, the first steps must be to establish a sectoral, regional, national, or international regulatory cap to restrict the cumulative volume of authorized GHG emissions, and then to require regulated pollution sources to obtain a permit containing GHG allowances for each covered source within the aggregate cap level set for the applicable target dates. One of the most controversial cap-and-trade issues is to what extent allowances should be given to grandfathered heavily-polluting industries, such as coal-burning power plants, and to what extent the polluters must purchase allowances through some kind of auction. Among the many critics, Harvard economist Martin Feldstein disparaged the revised Waxman–Markey Bill for its agreement to "give away some 85 percent of the permits over the next 20 years to various businesses instead of selling them at auction," including allocating "30 percent of the permits to local electricity distribution companies."[143] Dr. Feldstein concluded that the choice to give away most of the pollution allowances "only makes a bad idea worse."[144]

Individual firms may discharge GHGs only up to the limit of the allowances in their permits, and the cumulative discharges of all regulated sources cannot exceed the aggregate cap imposed. A given firm could retain or expand its discharges by purchasing more GHG allowances, but only when other GHG sources are able and willing to reduce their emissions by a comparable amount below the cap. Participating firms with low pollution control costs could choose to reduce their GHG emissions and acquire fewer allowances, or they could sell excess allowances to other polluters.

We must now ask: *Where will the discharges authorized by the GHG allowances go?* The allowance is a legal right to emit a designated amount of GHG pollution, usually one ton,[145] and there is nowhere for this pollution to go except into the atmosphere. The many discussions of cap-and-trade systems always employ a BAU baseline — typically the amount of GHG discharges in 2005 (Waxman–Markey and Kerry–Lieberman Bills) or in 1990 (Kyoto Protocol and EU ETS targets) — and they never really address what effect the discharges authorized by the cap-and-trade allowances will have on the cumulative atmospheric GHG concentration. The back-loaded emissions-reduction targets in the Waxman–Markey Bill and all other consensus emissions-reduction programs would allow equally as much GHG pollution to reach the atmosphere regardless of whether it is called persistent residual discharges or GHG pollution allowances. And the GHG discharges from cap-and-trade allowances, which represent legally authorized GHG emissions, will be just as persistent and harmful. The claimed "efficiency" of cap-and-trade systems appears to apply only to decreasing the pollution control costs of GHG dischargers, and not to reducing the increasing atmospheric GHG concentration and the dangers of global climate change.

Under the cap-and-trade system in the Waxman–Markey Bill, GHG polluters would have had to lower their emissions until they reached the limits specified by each of the five target-dates identified in Table 2 of Chapter II. Then they would not have had to reduce their discharges again until several years or two decades (2030–2050) in the future, depending on which target phase was the next one that must be met. The same back-loaded emissions-reduction pattern would apply if the Kerry–Lieberman Bill or some variant was eventually adopted by the Senate.

Table 3: Kerry–Lieberman Bill — Calendar Year Emissions Allowances ($MtCO_2e$)[146]

Year of Discharges	GHG Allowances	Year of Discharges	GHG Allowances
2013	4,722	2032	3,308
2014	4,635	2033	3,183
2015	4,548	2034	3,057
2016	5,524	2035	2,931
2017	5,417	2036	2,805
2018	5,310	2037	2,679
2019	5,202	2038	2,553
2020	5,095	2039	2,428
2021	4,941	2040	2,302
2022	4,788	2041	2,176
2023	4,634	2042	2,050
2024	4,481	2043	1,924
2025	4,327	2044	1,798
2026	4,174	2045	1,673
2027	4,021	2046	1,547
2028	3,867	2047	1,421
2029	3,714	2048	1,295
2030	3,560	2049	1,169
2031	3,434	2050 and years thereafter	1,043

Table 3, with its figures drawn directly from the Kerry–Lieberman Bill, shows the large volume of authorized GHG allowances intended for the years up to 2050 and beyond. The reason for the apparently sudden billion-ton increase in allowances in 2016 is that only the electricity and transportation sectors were covered in the first few years, but in 2016 all major industrial dischargers were included under the regulatory cap. However, from a climate-policy perspective, it hardly matters whether 4.548 billion tons of GHGs would be discharged in 2015 consistent with the number of allowances available, or whether 5.524 billion tons of GHG allowances and accompanying emissions would be authorized in 2016. In both years, these very large amounts of GHG emissions would be added to the already-too-high concentration of GHGs in the atmosphere, and the result would be a commensurate increase in climate change risks and present harms. The annual GHG emissions reductions in the Kerry–Lieberman Bill that would have been imposed if it had been adopted by Congress essentially ignored the vast "stock" of harmful GHGs already in the atmosphere. This is a classic example of the "reducing the increases" problem under which the annual allowances authorizing commensurate discharges would have steadily increased the cumulative GHG concentration in the atmosphere for at least the initial four decades of this counter-productive economic incentives program.

For example, in 2030 the discharge of roughly 3.5 billion tons of GHGs will be authorized by a comparable number of allowances. All of the approved GHG pollution — the emissions authorized by the cap limits and accompanied by valid allowances or offsets — would be discharged into the air, raising the atmospheric GHG concentration in exactly the same way that persistent residual emissions would exacerbate the greenhouse effect under a direct regulation scheme. I can see no climate-related progress or environmental benefits from using a cap-and-trade system in comparison to direct regulation. Advocates of cap-and-trade mechanisms might claim that when it enables lower-cost pollution control by various GHG sources, society will be able to afford a broader range of pollution control efforts. This argument is highly speculative because the regulated pollution sources, rather than climate change victims, would be much more likely to capture the resulting cost savings. In reality, most GHG sources have little incentive to go beyond the emissions-reduction program's mandated caps and annual targets.

The central conclusion of this analysis is that the same destructive GHG emissions will degrade atmospheric conditions whether the emissions originate as persistent residual discharges or are authorized by GHG allowances. The pollutants will be just as persistent under either form of regulation, and cap-and-trade programs with back-loaded emissions-reduction targets will be just as ineffectual and harmful as other "reducing the increases" regulatory programs would be. As long as GHG cap limits and target compliance dates are the same under a cap-and-trade system in comparison to a direct regulation approach, the same amount of harmful emissions will be added to the atmospheric GHG concentration in the same timeframe and will result in the same magnitude of climate change harms.

There are, however, a few reasons to conclude that a cap-and-trade scheme might be even less effective than a direct regulation approach from the perspective of their impacts on the atmospheric GHG concentration. Pollution control expenses may be significant for some types of businesses, but there are usually many other financial and operational factors of at least equal importance, including changes in consumer demand, opportunities for more efficient industrial processes, competition from domestic and foreign firms, the availability and cost of capital, and the impacts of economy-wide business cycles (e.g., recessions or inflation). Only in rare cases would the "tail" (GHG pollution control costs) wag the "dog" of large polluting enterprises with many other economic factors to balance.

What will be the effect on GHG discharges if a firm must go out of business, or if it must change its product line or production processes in ways that will generate fewer GHGs? Under direct regulation, a firm that chooses to close down or to modify its operations will eliminate all or some of the GHG pollution from its operations (not the harms from its past pollution). When a discharger must cease operations or convert to a GHG-free or low-GHG process, the result is that its GHG emissions will be reduced for purely business reasons. Under a GHG tax system, the same pollution reduction result will occur because no firm would continue paying a pollution tax when it has no further reason to continue discharging GHGs.

In contrast, under a cap-and-trade system, a closing or evolving business will often be able to sell GHG allowances it no longer needs to other firms that will be able to put out more GHG discharges because they have acquired more allowances. In effect, the GHG allowances may survive after the firm that initially obtained them has eliminated its need for them or has gone out of business entirely. This concern becomes greater depending on how long the lifespan of the allowance is: firms would seldom trade for an allowance with less than one year of remaining value, but allowances that authorize GHG discharges for several years or decades will be valuable and worth acquiring after the originating firm no longer requires them.

Some advocates of cap-and-trade mechanisms have argued for the establishment of GHG allowances "banks" that would permit firms to capture the future value of their allowances even if they no longer require them for their business operations.[147] For example, allowances banking and offset banking were included in the Kerry–Lieberman Bill and likely would be included again if similar legislation is eventually adopted.[148] Any regulatory system that encourages retaining allowances and corresponding GHG emissions when they are no longer needed for the business purposes of the firm that obtained them — or that creates a separate speculative business in trading allowances — is likely to increase the amount of GHGs that will be discharged into the atmosphere during each phase of the emissions-reduction program. Cap-and-trade is the only regulatory strategy that turns the legal right to put out more GHG emissions than necessary for particular business functions into a potentially valuable asset that would allow some firms to increase their allowable discharges when other firms reduce their need to pollute.

Dr. Hansen has made a somewhat similar argument: "[I]n a cap-and-trade world, acts of individual virtue do not contribute to social goals. If you choose to drive a hybrid car or buy a house with a small carbon footprint, all you are doing is freeing up emissions permits for someone else, which means that you have done nothing to reduce the threat of climate change."[149] He added: "Congress is pretending that the cap is not a tax, so it must try to keep the cap's impact on fuel costs small. Therefore, the impact of cap-and-trade on people's spending decisions will be small, so necessarily it will have little effect on carbon emissions."[150]

It is, however, unlikely that any politically tenable cap-and-trade system would require individual consumers to obtain GHG allowances commensurate with their activities or would give them the opportunity to trade personal allowances to other sources. Consequently, the Hansen criticism based on "freeing up emissions permits for someone else" is far more likely to occur in a business-closure context than in a consumer-altruism context.[151]

Dr. Hansen emphasized that cap-and-trade systems are more complicated than the carbon fee or tax system he prefers.[152] This assessment is similar to the remarks of Professor Roberta Mann, who argued that pollution taxes are preferable because "the devil is in the details" and a "cap-and-trade system has a lot more details than a carbon tax."[153]

In various passages of his recent book, Dr. Hansen argued that cap-and-trade proposals have proven especially vulnerable to the manipulations of special-interest lobbyists who spend very large amounts of money to control our political system, and he sharply criticized politicians willing to make deals with the lobbyists for their own personal benefit or to get "something done."[154] Hansen observed that if Wall Street firms, such as Goldman Sachs, are allowed to trade GHG allowances and to issue derivatives of the GHG allowances, which was explicitly authorized in the Waxman–Markey Bill, their "profits would be added to the fuel price" and ultimately the public would have to pay for this speculation.[155]

Dr. Hansen also explained how the virtues of the acid rain cap-and-trade program, which is always cited by market-forces proponents to show the effectiveness of a cap-and-trade system, are a "myth" because the "main solution was use of low-sulfur coal" which has no relevance for controlling and reducing GHG emissions.[156] He contended that the "horse-trading that made coal companies and utilities willing to allow this cap-and-trade solution did enormous long-term damage" because inefficient, obsolete coal-burning power plants were "grandfathered" under the Clean Air Act and allowed to continue their heavy air pollution for decades.[157] Dr. Hansen stressed that, despite the acid rain cap-and-trade system or perhaps because of it, "in 2009, there are still 145 operating coal-fired power plants in the United States that were constructed before 1950."[158]

The idea of "putting a price on carbon" has become more a catechism than a foundation for thoughtful analysis. Any serious assessment must examine the impacts of carbon prices in energy markets and other sectoral markets that generate a large amount of GHG pollution in light of the disparate endowments and political influence of competing market actors. If a cap-and-trade market for GHG allowances is implemented, the fossil fuel industries with trillions of dollars in assets, large government subsidies, and a history of strong political influence will surely have a great advantage over less-entrenched energy suppliers, such as renewable energy producers. It is difficult to understand how "putting a price on carbon" that arises from trading GHG allowances and carbon offsets under a cap-and-trade system would be able to overcome the massive inequality in wealth and power of the major GHG-polluting companies when compared to the recently established "clean technology" enterprises. It is even harder to believe that the "clean" industries will be able to overcome "dirty" fossil fuel industries in a carbon market competition within a sufficiently rapid timeframe to achieve notable reductions in climate change dangers.

In May 2011, there was a congressional debate over whether favorable tax treatments enjoyed by oil companies should be eliminated to decrease the federal budget deficit and to promote renewable energy initiatives. The congressional sponsors of one bill contended that the major "oil companies have made $1 trillion in profits over the last decade."[159] Assuming this estimate is reasonably accurate, how will "putting a price on carbon" under a cap-and-trade system enable "clean" energy producers to overcome the enormous imbalance in fossil fuel industry wealth and political clout? The failure of this oil-tax bill to obtain majority support in the US Congress despite the current emphatic focus on deficit reduction and energy independence is a clear indication that the political influence of the leading fossil fuel companies matches their great wealth.[160]

If the market price on carbon emissions resulting from a cap-and-trade scheme or carbon tax system is a low one, which has been true of every operational economic incentives program thus far, this market price is not likely to create strong enough incentives for "dirty" polluting firms to reverse the continuing growth of their GHG emissions. In practice, the low carbon price will undoubtedly take a long time, if not forever, to overcome the entrenched economic and political advantages of the fossil fuel industries and fossil fuel-producing countries. Putting any market price on carbon discharges will always create some kind of market-based incentive for carbon users to *consider* reducing their discharges, but there is no reason to believe that a low market price will create sufficiently compelling market pressures for fossil fuel dischargers and other major GHG polluters to stop relying on harmful but profitable practices.

On the other hand, imposing a high carbon market price, which has never actually been done, would raise the specter of severe social dislocation in many regions. It is hard to imagine a political system that would tolerate unrestricted high prices that might force an area's largest employers and most innovative firms to leave the area or to go out of business. A high carbon price could threaten the continuation of important social services such as ambulance operators, waste-treatment facilities, private universities and schools, and various government agencies. The likelihood of many political representatives abdicating their own "power of the purse" while allowing high carbon market prices to change the economic conditions and social welfare of their constituents is not very high, to say the least.

In June 2011, as one illustration, the EPA decided to delay issuing a major rule to restrict GHG emissions from large power plants as a consequence of sharp criticisms by Republicans in Congress and pro-business lobbyists, who argued that the "rule would have a profound effect on the price, supply and reliability of electricity by forcing modifications to, or the shutdown of, dozens of older power plants."[161] This antagonistic political position was adopted before any cap-and-trade system was actually launched or any specific emissions-reduction regulations were promulgated, and it is easy to imagine how much stronger the critical attacks would be if high market prices from an economic incentives program were already creating the projected social-dislocation effects.

Setting a market price for carbon that could achieve climate change mitigation goals is a far more complex process than the advocates of "putting a price on carbon" have acknowledged. As another example, the largest source of GHG emissions in the US is discharges from coal-fired power plants. These "dirty" GHG polluters are regulated by state utility commissions that must approve the electricity rates the companies can charge the public, and the utility commissions will often authorize revenue increases (higher consumer prices) to cover higher operating costs. Rate-setting is a quintessentially political process that will differ markedly among various states.

If a state utility commission chooses to allow a power plant company to recoup the cap-and-trade system costs of purchasing GHG allowances, or the cost of any carbon tax it must pay, there is no apparent reason why the carbon market price would lead to any meaningful reduction in harmful GHG emissions from the regulated firms. Having 50 different cost-compensation policies established by 50 state utility commissions in the US would lead to continuing friction and legal challenges based on claims of unfair or unequal treatments imposed on the electric utility companies.

In the same vein, GHG-polluting industries would often be able to "pass on their costs" to consumers, who will have little choice except to pay the higher prices resulting from economic incentive programs. Economists call the relationship between the price of goods and the extent of demand "price elasticity," which means in simple terms that for some goods a price increase (or decrease) will have a substantial effect on the volume of demand, but for other goods, such as electrical energy, a price change within realistic limits will be less likely to affect the market demand significantly. The price elasticity factor means that an economic incentives program may be relatively effective or ineffective in reducing GHG emissions, depending on the particular characteristics of the goods and services associated with GHG-polluting production activities in diverse areas.

Any "putting a price on carbon" advocate cannot presume that market incentives created by cap-and-trade systems, carbon offsets, or carbon taxes will have effective emissions-reduction results without carefully examining the specific conditions and probable effects in many different sectoral and regional contexts. The positive literature on cap-and-trade contains a great deal of happy generalization with a lot less evaluation of many disparate specific circumstances.

Many complex issues and pervasive uncertainties must be addressed before "putting a price on carbon" could attain the essential climate change mitigation. Instead of comprehensive analyses of the numerous problems, however, the economic incentives literature largely offers doctrinaire assurances of how wonderful carbon markets are in comparison to restrictive government regulation. In my opinion, the current literature and limited empirical evidence do not show that cap-and-trade systems or other economic incentives programs have any realistic probability of reducing climate change risks to an appreciable degree or that they would function any better than direct regulation mandates.

The "efficiency" arguments offered on behalf of cap-and-trade mechanisms contend that many firms would engage in research and innovation to cut their pollution levels in order to reduce their need to purchase GHG allowances or to allow them to sell their allowances to firms with higher pollution control costs. From the perspective of the atmospheric GHG concentration, it does not matter if firm A puts out 1,000 tons of GHGs or if it trades the allowances to firm B, which will then put out an additional 1,000 tons of pollution because it has acquired the appropriate number of allowances. This trade presumably will make money or save money for firm A, but it would do nothing to improve the atmospheric conditions or limit climate change harms because the traded GHG allowances would authorize continuing residual pollution that will progressively increase the GHG concentration in the air.

Although marginally different cap-and-trade models have been advocated by various proponents,[162] this economic incentives preference among the majority of environmental economists presumes that there is one "most efficient" pollution control approach no matter how deep (or shallow) the GHG pollution cuts and regulatory caps must be. I question this one-size-fits-all treatment and believe that the most efficient method for achieving a 30% or 50% GHG emissions cutback is not necessarily the same method that will be best for achieving an 80% or 90% GHG reduction.

A cap-and-trade program probably cannot set the aggregate GHG cap close to zero allowable emissions, or the claimed innovation and flexible trading benefits of economic incentive mechanisms would be unlikely to materialize.[163] The incentive for firms to participate in a cap-and-trade program comes from their ability to choose whether to purchase additional GHG allowances or to develop less costly pollution control methods that will reduce their need for allowances. Few economic benefits would stem from this market-based trading system if a strict regulatory cap authorizes only a very small number of allowances and all firms must therefore come close to phasing out their GHG pollution.

There is a limit to how much a business will invest in innovation to create better GHG pollution controls when very little permissible GHG pollution remains authorized — a draconian regulatory cap — with a comparably limited market for tradable GHG allowances. Under these conditions, most firms will be trying to create innovative GHG-free alternative technologies to meet the strict cutback standards instead of devoting major R&D efforts to improving pollution control methods for the limited remaining volume of GHG discharges. I have not found a single thoughtful analysis of the benefits and liabilities of cap-and-trade systems under conditions in which the regulatory cap must be set at a very stringent level, and yet the need for drastic GHG reductions verging on near-elimination is increasingly accepted by many climate scientists.[164]

The political and administrative designers of cap-and-trade programs are faced with a dilemma: If the regulatory cap is set close to zero GHG emissions, there will be few economic incentives to promote pollution control innovation and flexibility, which are the asserted market-based benefits of cap-and-trade systems. On the other hand, if the cap is set substantially above zero emissions, the program will become another "reducing the increases" failure that authorizes large volumes of persistent residual discharges leading to progressively higher atmospheric GHG concentrations.

We could design a cap-and-trade program in which the allowances shrink by a specified percentage every year to reduce the amount of GHG discharges over the life of the program. Yet, the faster and further the GHG pollution cap is decreased, the fewer cap-and-trade benefits are likely to accrue for participating dischargers. And the slower the GHG cap is reduced, the more persistent residual GHGs will be discharged into the atmosphere, compounding the greenhouse effect and climate change. It is important to stress that a 50% cap or 80% cap established by a cap-and-trade program, exactly the same as a 50% cap or 80% cap imposed by direct regulation, would authorize the discharge of large quantities of persistent residual GHGs that will increase the cumulative atmospheric GHG concentration every year.

The most extensive evidence on the performance of a large-scale cap-and-trade program comes from the first two phases of the EU Emissions Trading Scheme (ETS), which EU officials maintain have been of significant educational value despite the total absence of success in significantly reducing European GHG discharges or limiting climate change risks. The literature on ETS mistakes and potential corrective measures identifies numerous serious problems created by the system's design and implementation choices.[165] These problems include over-allocation (too many allowances issued for the chosen regulatory cap); leakage (regulated pollution can shift to an unregulated location with no actual net emissions reduction); intense political lobbying over the chosen cap and allocation method for the distribution of allowances (a maximum of 5% were sold at auction, while the rest of the allowances were given free to many grandfathered polluting firms); conflicts and competition among participating jurisdictions; competitive disadvantages for EU regions that imposed relatively high pollution control standards and a correspondingly strict cap; limited information on the annual volume of GHG discharges from thousands of pollution sources; lax or ineffective monitoring and enforcement efforts; price volatility and private-interest manipulations in the budding markets for ETS allowances; and billion-dollar windfalls for utilities that were given free allowances but passed the opportunity-costs of the allowances onto consumers anyway.[166]

The revised Waxman–Markey Bill was on its way to duplicating many of the worst ETS mistakes before the Senate refused to accept this cap-and-trade approach. Under the amended Waxman–Markey Bill, at least 80% of the allowances would have been given away for free to many of the largest GHG polluters in the nation, and the bill would not have prevented these polluting firms, especially power companies, from obtaining the same kinds of "windfall profits" that occurred from similar grandfathering under the ETS.[167]

In response to pressures from coal-mining states, mine workers, and energy businesses, the Waxman–Markey and Kerry–Lieberman Bills would have continued to offer subsidies and regulatory exclusions for coal and other fossil fuels that are at the core of US greenhouse gas emissions policies. As predicted by many critics, the revised bills would have enabled speculators and other private actors in the "carbon market" to focus on making profits from GHG allowances trading, rather than on limiting climate change dangers. The adage that "those who fail to learn from history are compelled to repeat it" appears wholly applicable here, but the inadequacies of these bills arose mainly from the weaknesses of our political system and the primacy of special-interest lobbying over the needs of national and international human welfare. Sadly, even if we do learn to recognize these selfish, short-sighted private-interest pressures, this does not mean that environmentalists will ever be strong enough politically and economically to overcome them.

The critical point here is that the emissions-reduction provisions of the Waxman–Markey and Kerry–Lieberman Bills, and similar legislative and Administration proposals with cap-and-trade provisions, cannot succeed any more than direct regulation could in attaining demonstrable climate change progress because they would fail to stabilize or decrease the accumulation of greenhouse gases in the atmosphere. No regulatory approach, whether direct regulation or economic incentives measures, can produce real climate change benefits as long as it allows the concentration of persistent GHGs in the air to grow higher.

Even if all of the identified EU ETS problems and congressional bill problems could be effectively fixed, which is implausible, the cap-and-trade programs would still be inadequate to control the cumulative build-up of harmful GHGs in the atmosphere. Reducing pollution control costs, which is desirable as long as the GHG pollution in the air is effectively restricted, does not show that market-based economic incentive programs will perform any better at curtailing climate change risks than direct regulation would, because cap-and-trade systems have not been designed to overcome the continuing accumulation of persistent residual discharges and the resulting growth of the atmospheric GHG concentration. In short, the major focus of cap-and-trade programs is reducing the costs of GHG emissions reduction. These programs have not been focused on the need to reduce current and future concentrations of GHGs in the air, which is the crucial requirement for overcoming climate change hazards.

Carbon Offset Programs

Carbon offset programs are a means to exchange a number of tons of GHG pollution reduction for an equal amount of pollution continuation. Various offset programs allow program participants with relatively low pollution control costs to profit by cutting their GHG discharges and selling the resulting emissions reductions as offset credits that could be subtracted from the buyers' pollution control requirements.[168] The buyer will have to pay an offset credit price — usually set by trading in a carbon market — to avoid eliminating a selected amount of its own discharges, while the offset seller must cut its emissions by the corresponding volume of GHG emission reductions at a presumably lower per-ton pollution control cost.[169]

In a carbon offset market, the emissions-reduction reality is that a given amount of GHG pollution control anywhere can be just as effective or ineffective in abating climate change impacts as a comparable reduction anywhere else. This is true because GHG emissions are fungible and are mixed in the atmosphere over time regardless of where they were initially discharged. Offset programs can assist GHG-polluting nations, corporations, or consumers to reduce net emissions-reduction costs by letting them purchase lower-cost offset credits. In addition to shifting some expenses of GHG emissions reduction to a lower-cost basis, offset programs may also assuage consumer guilt for unsustainable behavioral choices[170] and may help improve the public relations images of businesses or countries that are trying to "declare themselves carbon neutral" by purchasing enough offset credits to equal their cumulative GHG emissions.[171]

The critical problem with these schemes is that the offset credit buyer will be entitled to continue pumping substantial residual GHGs into the atmosphere on the rationale that these discharges would be counter-balanced (offset) by the permanent, reliable, verifiable GHG emissions reductions offered by offset sellers. In other words, the amount of GHG pollution authorized by offset credit purchases is the equivalent of persistent residual discharges or GHG allowances. Carbon offset programs appear to be mechanisms for maintaining an allowable level of GHG pollution discharged annually into the air, but the offsets are seldom, if ever, satisfactory means for reducing the atmospheric GHG concentration. Many implementation and enforcement problems with carbon offset programs present the same types of systemic weaknesses as cap-and-trade systems, but each kind of offset program has distinctive characteristics and vulnerable points.

Here is a brief summary of three kinds of offset programs: voluntary consumer-oriented offsets; carbon offsets intended to reduce the costs of national emissions-reduction requirements or to substitute for GHG allowances in national cap-and-trade systems; and international offset programs, such as the UN Clean Development Mechanism, in which the offsets are produced in developing countries and sold to polluters in more affluent nations. National and international carbon offset programs often provide "co-benefits," such as community development assistance or biodiversity protection, which in practice may be equally or more important to the offset-originating countries than lower-cost emissions-reduction savings would be. As a generalization, under carbon offset programs the offset suppliers benefit by receiving money or development assistance in return for the emissions reductions they provide; the offset buyers benefit by obtaining pollution control reductions at a lower cost than would be required for them to cut their own discharges; and the atmospheric GHG conditions will benefit very little, if at all.

Voluntary Offset Programs

As examples of voluntary offset programs, Marriott Hotels has adopted a plan that enables altruistic customers to pay for offsets intended to compensate for, or neutralize, GHGs that may adversely affect the Brazilian Amazon region.[172] A number of major airlines have created offset programs that will allow environmentally-disposed customers to pay to some extent for the GHG pollution harms resulting from their air travel activities.[173] Following a UN Environment Programme (UNEP) initiative to reduce the carbon impacts of the 2010 World Cup, the sportswear manufacturer, PUMA, promised to "offset the CO_2 footprint of PUMA-sponsored national football teams taking part in the Football World Cup this summer in South Africa — a total of 336 players and officials."[174] The common theme with these offset credits is that they are completely voluntary and are not produced, purchased, or scrutinized by any kind of official regulatory regime.

In my opinion, voluntary consumer-oriented offset programs are another manifestation of the "every little bit helps" fallacy. Very few airline and hotel customers are purchasing offset credits that would come close to equaling the GHGs their flights or stays generate. This means that a very large amount, almost certainly more than 95%, of the GHGs associated with these activities will constitute persistent residual discharges that increase the atmospheric GHG level and enable climate change risks to grow worse. The practical consequence of voluntary offset programs is to cut worldwide GHG emissions by a miniscule amount, thereby only "reducing the increases" in atmospheric GHG concentrations without restraining the ongoing growth in GHG pollution. The only arguable value of voluntary carbon offsets is that they will reduce some of the costs of decreasing GHG emissions by a very small amount that is supposedly "better than nothing," though they will ultimately achieve no appreciable climate change benefits.

In contrast to the cost-saving claims of offset proponents, the availability of voluntary offset programs may encourage offset credit buyers to engage in more frequent polluting activities than they otherwise would on the rationale that purchasing offsets will counter-balance the climate degradation from their activities. One well-publicized example is the large amount of energy consumed by former Vice-President Al Gore's farm in Tennessee, estimated at about 20 times the energy consumption of typical households.[175] Gore excused his high energy usage on the grounds that he purchased enough offset credits to make his farm's energy consumption "carbon neutral." Yet, without the carbon offset program availability, he would have had more incentive to reduce his cumulative energy usage, rather than shifting the pollution control offset locus to other parties in other places.

Voluntary offset programs normally entail difficult "free rider" problems. On a general plane, the *non-payers* (people and businesses that neither buy offsets nor reduce their GHG discharges) cannot be prevented from enjoying whatever climate benefits may accrue from the efforts of the *payers* who are trying to improve social or environmental conditions. In effect, the non-payers or free-riders can use their wealth to satisfy their personal preferences or business needs while they nonetheless cannot be excluded from any decreased harms or risks that would accrue if voluntary *payers* are able to achieve some form of climate improvement. Under these circumstances, many people will choose not to pay for mitigation or remedial costs because they cannot be excluded from enjoying the resulting climate benefits, if there are any.

Rather than imposing mandatory regulatory initiatives that require the costs of climate change damages to be incorporated or "internalized" into the prices of the enterprises that generate harmful GHGs, the voluntary offset programs allow the great majority of people and businesses to avoid paying for the climate damages caused by their individual or shared activities. Some altruistic people will contribute some amount of money to voluntary mitigation efforts because they want to feel they are "part of the solution," but this funding will doubtless be limited by the obvious unfairness of free-riding behaviors and will remain far below the amount necessary to achieve meaningful climate improvements.

Nearly all environmental experts agree that voluntary efforts, including purchases of voluntary carbon offsets, cannot effectively overcome the collective-action problems associated with climate change risks.[176] Yet, the advocates of voluntary offset programs continue to proselytize consumers and businesses to purchase carbon offsets that supposedly will somehow help combat climate change hazards.[177] Instead, these ineffectual, often expensive voluntary efforts are no more likely to improve climate conditions than to degrade them, and the case for reliance on voluntary carbon offsets certainly has not been demonstrated persuasively.

In a report on the condition of voluntary carbon markets in 2011, the authors found that: "In 2010, the volume-weighted average price of credits transacted on the voluntary OTC [over-the-counter] market fell slightly to \$6/tCO_2e from \$6.5/tCO_2e in 2009."[178] The report multiplied this \$6 average offset price by their estimate that a total of 98 million offset credits were traded in 2010, leading to their calculation of an aggregate carbon offset market value of about \$415 million.[179] This substantial annual sum only indicates how much money changed hands and how much profit might have been reaped, not how much mitigation progress was achieved by these transactions. The buyers will use the voluntary carbon offsets to justify putting out a comparable amount of discharges that will be exactly the same as persistent residual discharges and will similarly increase the atmospheric GHG concentration. The \$6 per ton offset price will likely create only a negligible mitigation incentive, if any, though it may occasionally inspire a limited amount of pollution-control innovation and emissions reductions by potential offset sellers. I am not contending that the voluntary offset market is completely useless, but its diversions of time, money, and other resources from more effective mitigation approaches will almost certainly outweigh any minimal climate benefits this type of program could attain.

The key point is that every ton of GHG discharges the offset producers are able to eliminate will enable a comparable emissions increase by the offset buyers. For example, shifting 98 million tons of allowable GHG discharges from the offset sellers to the offset buyers certainly will not contribute materially to reducing the greenhouse effect and related climate change harms. Moreover, people who normally would not engage in a specific polluting activity may choose to undertake that activity because they ostensibly can absolve their harmful actions by purchasing a comparable number of carbon offsets. It is thus very possible that voluntary carbon offsets may have the negative effect of increasing carbon discharges rather than reducing them.

Offsets as Part of National Emissions-Reduction Programs

The Waxman–Markey[180] and Kerry–Lieberman Bills[181] included very elaborate offset provisions that would have enabled carbon offsets to become potential substitutes for a substantial percentage of the direct regulation permits or cap-and-trade allowances. The offset credits are meant to reduce the cost of GHG emissions reduction in the same hypothetical way that tradable allowances would, and to make the emissions-reduction process more palatable for diverse interest groups that will be affected by the legislation's energy efficiency and climate change provisions. Unlike voluntary consumer-oriented offsets, these national or regional offset programs are tied to regulatory requirements that dictate the types and numbers of valid offsets as well as the necessary implementation steps for individual offset credits, which must be overseen in detail by the applicable regulatory agencies. Because government-approved offset credits can replace mandatory pollution control permits or GHG allowances, the nationally-accepted offset credits must possess a high degree of reliability and credibility.

Each approved offset credit would authorize a ton of GHG pollution to be discharged into the atmosphere. If greenhouse gas producers or users can purchase a billion offsets at lower cost in comparison to their own pollution control expenses, this "efficiency" nonetheless means that a corresponding billion tons of persistent residual GHGs will continue to be discharged into the atmosphere under the national offset program. It is easy to identify the claimed fiscal savings from reliance on carbon offsets, but there is no similar climate change benefit from allowing large quantities of GHG discharges to increase the atmospheric GHG concentration every year. The carbon offset credits entail substitute pollution, not reduced pollution.

It is doubtful whether carbon offset programs in practice will really reduce the aggregate costs of cutting GHG emissions after all the implementation, compliance, enforcement, and disclosure costs of necessary government and business activities are counted. The Kerry–Lieberman Bill, for example, contained nearly 1,000 pages of detailed provisions, including more than 100 pages devoted to national and international carbon (and carbon-equivalent) offsets.[182] The complex implementation and oversight requirements associated with operating these offset measures "efficiently" would require a large number of administrative decisions and offset producer choices that are likely to become regulatory nightmares in practice.

As one illustration of the operational complexity of many GHG carbon offset programs, the Kerry–Lieberman Bill authorized a range of "offset projects relating to emission reductions from domestic agriculture and forestry."[183] Before farmers or foresters could take advantage of these sectoral offset projects, the Secretary of the Department of Agriculture would have had to:

(1) gather inventory data on carbon stocks and fluxes to inform rulemaking with respect to the agricultural and forestry sectors;

(2) administer as the lead agency the duties prescribed under sections 734, 735, 736, and 739 for agricultural and forestry offset projects, in consultation and coordination with other relevant agencies;

(3) prepare the Forest Service, the Natural Resources Conservation Service, the Farm Service Agency, and other relevant entities to make available to landowners and offset project representatives carbon sequestration data and other information on agricultural and forest land that are necessary to assist landowners and project representatives in estimating carbon sequestration rates by land area or appropriate region, forest type, soil type, and other appropriate factors;

(4) make available technical assistance to landowners undertaking activities for the generation and sale of offset credits derived

from activities on the land of the landowners, including information about working with aggregators and third-party verifiers pursuant to section 737;

(5) take into consideration expanding existing training and accreditation programs of the Natural Resources Conservation Service for third-party technical service providers to provide training and accreditation for third-party verifiers pursuant to section 737;

(6) conduct, as appropriate, outreach, education, and training through the extension services of land-grant colleges and universities; and

(7) promulgate such additional regulations as are necessary to carry out the functions of the Secretary under this part.[184]

This is only one set of convoluted administrative requirements under the complicated national and international offset provisions in recent congressional bills, which would have required the promulgation of hundreds of regulatory rules and standards that subsequently must be applied to thousands of individualized agency determinations on the creation and monitoring of specific offset credits. As has been the case with many US pollution control programs in previous decades, carbon offset rules and specific individualized determinations will be challenged by business and environmental interest groups on the rationales that they are arguably too strict or arguably too lenient.

For every authorized carbon offset, up to the maximum limit of two billion national and international offsets, the Kerry–Lieberman Bill provided that "the appropriate official [must] establish standardized methodologies for: determining additionality; establishing activity baselines; measuring performance; and accounting for and mitigating potential leakage."[185] It is difficult to predict how many years or decades this process would take before it could be implemented in a reasonably effective manner, or how expensive the long-term implementation and enforcement requirements would prove. Agricultural and forest conditions vary frequently, especially as a result of increasing climate change, which means that a satisfactory offset practice one year may become an unsatisfactory failure the next.

I have found no comprehensive attempt to determine the likely consequences of the many diverse implementation problems and delays associated with the proposed offset programs and other complex regulatory provisions that would influence the residual GHG discharges and annual atmospheric GHG concentrations. In the presence of idealistic wishful thinking, the economic incentives advocates promoting cap-and-trade and carbon offset systems may not have recognized that the great complexity and high implementation costs of these mechanisms have been among the factors contributing to the reluctance of Congress to adopt a final climate change and energy efficiency bill into law.[186]

Rather remarkably, the carbon offset plans in recent legislative proposals virtually ignore the impacts of climate change on the variability of the weather conditions affecting agricultural and forestry processes. The offset credit programs have aimed to reduce the costs of emissions-reduction programs by eliminating a considerable proportion of industrial pollution control and replacing it with less expensive agricultural- and forestry-based GHG reductions. It is easy to understand that the sponsors of this legislation included the offset provisions to gain political support by pleasing or placating American agricultural and forestry interests. However, effective climate change mitigation measures must be able to operate under actual climate change conditions. Because of the increasing variability and intensity of climate change impacts, greater reliance on agricultural offsets would shift a considerable proportion of the responsibility for cutting greenhouse gases from industrial pollution sectors to less predictable, controllable, and reliable agricultural and forestry sector operations that are subject to extreme weather conditions, weather unpredictability, and harmful ecological transformations.

In recent years, for example, the southern, southwestern, and western regions of the United States have suffered their worst droughts in the past 50 to 100 years.[187] One scientific research study concluded: "The effects of drought on vegetation under warmer conditions can be severe, as highlighted by recent regional-scale woody-plant die-off across the southwestern United States and around the globe.

Worldwide, many coniferous tree species are experiencing widespread, historically unprecedented mortality, mainly as a result of drought and the eruption of tree pests, such as bark beetles."[188]

During the same period, southern, midwestern, and southwestern states have suffered from unusually severe storms and flooding.[189] An unprecedented number of wildfires have damaged California and other western states.[190] In "the largest known insect infestation in the history of North America,"[191] the mountain pine beetle is destroying tens of millions of acres of forests in British Columbia and western American states, an ecological and economic disaster that is happening because global warming has moderated the frigid winter conditions that once protected the forests against the growth of pine beetle larvae.[192] The gradual warming of winter temperatures is also threatening the northeastern Sugar Maple tree, which is among the most economically important hardwood species.[193]

Other climate change risks that could adversely affect agricultural and forest productivity include heatwaves with temperatures over 100 degrees Fahrenheit[194]; extreme weather events such as more tornados[195] and hurricanes; and lower crop yields resulting from excessive heat and more frequent water supply shortages.[196] Because global warming and climate change are worldwide phenomena, the same kinds of extreme or unpredictable weather conditions may affect agricultural and forestry productivity in all nations and hence will be just as likely to undermine the reliability of international offset programs.[197] The point of this discussion is not to recapitulate the various dangers of climate change, but rather to show how unreliable carbon offsets relying on agricultural and forestry productivity are sure to be. What kind of enforcement action or fine will carbon offset program administrators try to impose if a hurricane, tornado, drought, flood, or heatwave wipes out most of the agricultural and forestry growth in a large region? I predict no enforcement in the presence of supposed "acts of God" and no means to prevent persistent GHGs from reaching the atmosphere in these contexts despite ostensible verification of many agricultural and forestry carbon offsets.

Whether offset projects involve attempts to create additional carbon sinks or to expand biomass production to facilitate alternative energy production, these efforts will inescapably depend on erratic and largely uncontrollable environmental conditions. Pouring many billions of dollars into agricultural and forestry offset programs — which only provide pollution substitutes, not permanent reductions or sequestration solutions — cannot make sense when compared against the utility of developing GHG-free replacement technologies that do not produce persistent residual GHG emissions and do not depend on optimum weather conditions that cannot continuously prevail while climate change hazards become steadily worse.

International Offsets and the Clean Development Mechanism

Some nations have a much better record of addressing climate change problems than the US has. For example, Norway recently announced that it will become the first "carbon-neutral" nation by 2030.[198] The required GHG reductions would come partly from large-scale renewable energy projects and partly from Norwegian purchases of Clean Development Mechanism (CDM) offsets and additional offset credits from other international programs.[199] Although Norway deserves praise for its renewable energy and alternative energy programs,[200] it is rather ironic that many of the international offset credits it is purchasing will be financed by revenues derived from Norway's North Sea oil production.[201]

The more troubling problem is that developed nations are purchasing international offset credits to avoid cutting their own GHG emissions as much as their self-imposed legal obligations would otherwise require. For example, one assessment of carbon tax and offset proposals noted that "in the case of Norway, emissions have actually increased by 43 percent per capita."[202] Yet Norway claims it is on the road to carbon neutrality. Even when the offset process represents a sincere attempt to reduce climate change harms, rather than mainly paying developing nations to help wealthy states avoid cutting their GHG pollution, the offset programs and similar North–South redistributive efforts are promoting "reducing the increases" GHG accumulations that ensure atmospheric GHG levels will continue to become worse.

The UN Clean Development Mechanism is the most widely known and frequently used international carbon offset program.[203] It was established in conjunction with the Kyoto Protocol to reduce the pollution control costs of participating developed member-states (Annex I nations) by allowing them to purchase offsets derived from emissions-reduction projects in developing nations. As a consequence, many CDM projects have produced "co-benefits" in which economic and social development payoffs obtained by the non-Annex I countries — the states not required under the Kyoto Protocol to make specific GHG cuts — are treated as equally or more important than any asserted climate change benefits. The CDM process and many individual CDM projects have been critically evaluated on grounds similar to the weaknesses of the cap-and-trade systems, including leakage, insufficient permanence, inadequate monitoring and disclosure, lax enforcement of offset standards, excessive speculation in carbon markets, corruption among government officials, and broadly ineffective environmental protection efforts in the developing nations.[204]

The tensions between the highest priorities of developed and developing nations, and between the goals of climate change mitigation and expanding economic growth, have greatly politicized the design and implementation problems of international offset programs.[205] One continuing dispute focuses on the extent to which affluent donor nations or carbon offset purchaser nations should be able to closely monitor implementation activities in developing states to ensure they are consistent with the purposes of the offset programs. Indeed, the concern for CDM implementation in developing countries has led to creating a set of international oversight standards known as "MRV," which means the need for "Monitoring, Reporting, and Verification" or sometimes "Measurement, Reporting, and Verification."[206] These essential implementation tasks are very difficult to accomplish in many developing nations where the "rule of law" is weak and the ownership of natural resources is often disputed.

A CDM project sponsor must show that additional GHG emissions will be reduced in comparison to the initial baseline design of the project — otherwise the investor or sponsor nation would be paying for a development project in another state without getting any GHG pollution reduction benefits. To explain the concept of *additionality*,[207] UN guidelines provide that a "CDM project activity is additional if anthropogenic emissions of greenhouse gases by sources are reduced below those that would have occurred in the absence of the registered CDM project activity."[208] In other words, any emissions cutback below the GHG volume that would have been discharged under a BAU project design would justify an administrative finding of "additionality" for the offset project, which in my view is far too minimal an emissions-reduction standard. The UN requirements for demonstrating project additionality, complying with MRV standards, and overcoming various other implementation and enforcement problems have given rise to many criticisms of the CDM process.[209]

From the perspective of this book's central theme, the most important concern is that CDM offset projects, even when they fully meet the additionality and MRV tests, are still "reducing the increases" emissions-reduction measures that cannot yield any actual climate change benefits. Instead, their continuing GHG discharges will contribute to making climate conditions worse every year. Consider this diagram drawn from UNEP documentation that purports to show a CDM offset project in compliance with the additionality requirement:

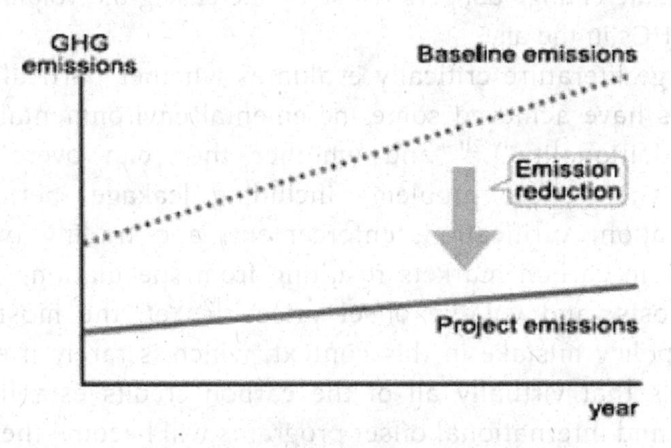

Figure 3: The "Additionality" Conception[210]

In Figure 3, the "additional" emissions reduction provided by the CDM offset project is represented by the differential between the

"Baseline emissions" dotted line and the lower "Project emissions" solid line. I doubt that many actual CDM projects have reduced their GHG discharges from the Baseline design level by more than 50%, as this Figure happily depicts. Nevertheless, the vital point is that the *combination* of the new "Project emissions" and the background GHG emissions shown by the horizontal "year" line leads to a higher cumulative GHG level that increases the atmospheric GHG concentration and degrades climate conditions.

Perhaps by coincidence, this "additionality" diagram is almost the same (except for the labels) as my "reducing the increases" image in Figure 1. In both instances, the annual GHG discharge level may be sharply reduced in comparison with the BAU level of GHG emissions, but the combination of the atmospheric "stock" of GHGs and the new "Project emissions" will create a cumulative increase in the atmospheric GHG concentration and greenhouse effect. The hypothetical project in Figure 3 might decrease the pollution control costs for the investor nation and it might provide desired co-benefits for the offset-producing developing country, but this CDM offset project would not help stabilize or decrease the atmospheric GHG concentration and climate change risks in any way. Indeed, this hypothetical carbon offset project would make climate change dangers worse by increasing the volume of persistent GHGs in the air.

A large literature critically evaluates whether particular offset programs have achieved some incremental environmental protection ("additionality"),[211] and whether they can overcome the familiar GHG offset problems including leakage, permanence, quantification, verification, enforcement, and trading incentives problems in carbon markets resulting from speculation, administrative costs, and volatile offset prices.[212] Yet, the most crucial climate-policy mistake in this context, which is rarely if ever discussed, is that virtually all of the carbon credits established by national and international offset programs will become the equivalent of persistent residual GHG emissions that are certain to make climate change more harmful when the greenhouse gases authorized by the offset credits are discharged into the atmosphere. Here is another illustration from UN public-interest documentation that clearly portrays this concern:

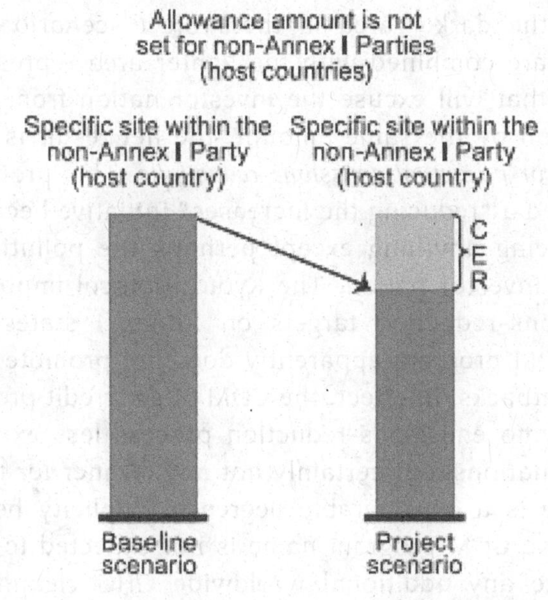

Figure 4: A CDM Project Generating CER Offset Credits[213]

As depicted in Figure 4, the GHG emissions from the "Baseline scenario" or business-as-planned project design would be reduced by the host developing nation in the "Project scenario" upon the receipt of development-oriented foreign aid or technology transfers from an affluent investor nation subject to emissions-reduction targets under the Kyoto Protocol. The tradable carbon offset credits acquired by the investor nation are called "Certified Emissions Reductions" (CERs), which is no more than the generic name for offset credits under the CDM program. The CERs are equal to the claimed GHG "reductions" reflected in the Project scenario, which would enable the investor nation to use the CERs to avoid reducing its own GHG discharges by the designated proportion of its emissions-reduction responsibilities. But what would happen to the dark area in the Project scenario column that represents ongoing GHG discharges? In both the Project scenario and the

Baseline scenario, the dark areas show persistent residual emissions that will increase the atmospheric GHG concentration.

It is important to recognize that when the project's GHG discharges — the darker area in the Project scenario column of Figure 4 — are combined with the lighter area representing the CER offsets that will excuse the investor nation from cutting its GHG pollution by the same amount, the net result is that there will be *little or no actual emissions reductions*. This project cannot even be called a "reducing the increases" initiative because it will not be reducing anything except perhaps the pollution control costs of the investor nation. The Kyoto Protocol imposes miniscule emissions-reduction targets on Annex I states-members, while the CDM program apparently does not promote additional emissions cutbacks. In effect, the CDM offset credit program may make the Kyoto emissions-reduction process less expensive for the affluent nations, but certainly not any cleaner for the atmosphere. There is a considerable degree of duplicity because the "Clean" in the CDM program name is not expected to, and does not, generate any additional worldwide GHG cleanliness. The international offset process is meant to achieve lower-cost pollution substitution, not significant pollution reduction.

The CDM conception is a reasonably generous way for affluent countries to pay for the "co-benefit" of international development assistance and to redistribute wealth to less wealthy nations. Yet, the asserted global warming benefits arise from wishful-thinking claims about "reducing" GHG emissions from projects that may never have been created in the first place without the availability of CDM international payments. Under the CDM system, developing host nations always have an incentive to exaggerate the Baseline scenario pollution levels in order to expand the asserted volume of emissions reductions and the CERs resulting from foreign CDM investments. At the same time, project administrators always have an incentive to overstate the volume of CERs that can be gained by investor nations because this will serve as a cost-saving offset against more of the GHG pollution the developed nations are committed to eliminate.

CDM projects can be located only in developing countries that normally have no national GHG emissions-reduction cap and also have no specific obligations under the Kyoto treaty. This means a host nation may reduce the GHG emissions from a particular project to create a corresponding number of tradable CERs while it continues to build hundreds of "dirty" coal-burning power plants further down the road or river. This is more than conjecture because China has been the leading host nation beneficiary of CDM projects[214] despite constructing more dirty coal-burning power plants at a more rapid rate than any other nation on Earth.[215] The net result is that CERs from CDM projects in China allow European Union investor nations and other Annex I members to reduce their emissions-reduction efforts, while China's rapidly increasing GHG discharges are essentially rendering all of the limited Kyoto Protocol pollution control targets meaningless.

The CDM international offset program is a compelling example of why "better than nothing" is not an acceptable climate-policy standard when mitigation programs fail to achieve tangible climate benefits. If the regulatory program relying on tradable carbon offsets cannot attain meaningful prevention or adaptation benefits under realistic climate conditions, this initiative would clearly be a waste of scarce resources and a threat to undermine potentially more effective approaches. I see no point in expending irretrievable amounts of money, time, and efforts on mitigation programs that do not offer any substantial climate change improvements even when they are working as well as they could possibly function. Unfortunately, many climate experts apparently believe that any reduction in GHG discharges is "better than nothing" without recognizing the opportunity-costs of wasted investments that will never overcome the climate change dangers humanity is confronting.

Carbon Taxes, Fees, or Charges

Carbon taxes, fees, and charges are synonyms for the same climate-policy instrument, which could serve a number of important purposes in promoting climate improvement efforts.[216] Greenhouse gas taxes, especially taxes on carbon dioxide emissions, could be used to raise revenues that will support mitigation, adaptation, and technological innovation programs. The carbon taxes could be imposed to deter or penalize environmentally harmful activities. The taxes could sometimes help correct market prices that have been distorted by externalities and misplaced subsidies. Selective user taxes could also help shift consumption choices toward less destructive activities.

The revenue-collection function of a carbon tax, fee, or charges program would be vital in the climate change context because expensive technological and behavioral alternatives must be available to achieve the necessary transformation to a low-GHG economy and society. It is equally important that international climate programs receive substantial financial resources to pay for GHG-free technology transfers and other forms of foreign assistance to developing nations that would enable them to curtail their GHG discharges without sacrificing increased economic development.

A tax on GHGs supposedly could be used to "level the market system playing field" by offsetting the effects of negative externalities and inefficient or inequitable subsidies. Indeed, pollution is the classic harmful externality because polluters normally do not bear a substantial proportion of the social costs of their activities. As Sir Nicholas Stern, the noted British economist and climate expert, observed: "Greenhouse gas (GHG) emissions are externalities and represent the biggest market failure the world has seen."[217] A carbon tax could go beyond merely "leveling" market system prices for competing enterprises by imposing a much higher price on fossil fuel uses in order to give "clean" energy competitors a significant market advantage.

A carbon tax could sometimes help deter destructive activities, such as discharging high levels of industrial GHG pollution or high levels of fossil fuel pollutants from motor vehicles, which contribute substantially to the growth of the atmospheric GHG concentration. This kind of deterrent "sin tax" could be combined with revenue-generating and market-leveling functions to promote several desirable purposes with one fiscal instrument. The function of a carbon tax in this context is the straightforward one of attempting to reduce harmful activities by increasing their market prices. The tax may also encourage greater investments in technological innovation by firms attempting to find less costly ways to supplant targeted "dirty" practices.

It is hard to envision how many, if any, renewable energy companies would be able to compete effectively against established fossil fuel companies as long as the fossil fuel producers do not have to pay for the worldwide harms they have been causing for many decades.[218] The fossil fuel industries and other GHG generators have benefited greatly from relatively low production costs and consumption prices because they have been heavily subsidized by low tax and royalty rates, inexpensive licenses for resource extractions on public lands, government research investments, and other de facto forms of public assistance.[219] Indeed, a 2010 study by Bloomberg New Energy Finance concluded that the fossil fuel industries are still receiving worldwide subsidies roughly 10 to 12 times larger than the government subsidies supporting renewable energy initiatives.[220]

Imposing a carbon tax on fossil fuel energy production could enable renewable energy mechanisms and other "clean" energy technologies to compete on more even market-price terms, or preferably on better than even terms, against the fossil fuel companies that have benefited from strong financial, political, and legal advantages during the past century. To a degree, the carbon tax could also serve as a substitute or supplement for meager law enforcement efforts that have usually imposed very small penalties on fossil fuel producers that should have been held responsible for many major damaging actions, such as severe oil spills and widespread water contamination from coal mining.[221]

Carbon taxes, fees, or charges are versatile instruments that could be aimed at funding different methodologies requiring substantial alterations to enable conversion to a GHG-free economy. For example, gradually increasing energy-efficiency taxes could be imposed on new appliances, motor vehicles, and buildings.[222] Pollution taxes, fees, or charges could be aimed at consumption activities that generate substantial GHGs, such as driving gas-guzzling cars and taking long-distance airline flights. Taxes could be aimed at increasing the cost and reducing the volume of harmful natural resources inputs, such as coal, oil, and natural gas, which generate GHG discharges.[223] Diverse pollution taxes and energy taxes could also be targeted at low-volume high-price luxury consumption products, such as motor-yachts and million-dollar mansions, or at high-volume low-price everyday products, such as plastic bags and recyclable beverage bottles.

The conception of using dedicated taxes to promote specific public-policy purposes is certainly not a new one. For example, the US federal government collected billions of dollars of automotive fuel taxes to pay for the Interstate Highway System, and imposed additional user fees on commercial trucks that increased the wear and maintenance costs of the highways. Tobacco and alcohol taxes have been repeatedly raised to help pay for public health disclosures about the dangers of smoking and drinking, and also to reduce the prevalence of these socially harmful practices by raising their prices.

Many societies have imposed various kinds of taxes or user fees to support the provision of bridges, national parks, museums, broadband Internet connections, and numerous other public goods. As another example, for more than a decade the federal government imposed a "Superfund" tax on the chemical industry to help clean up toxic waste sites. Then business-oriented politics and special-interest lobbying under the Reagan Administration triumphed over health, safety, and environmental protection concerns, which is precisely the same thing that has happened to congressional climate legislation in recent years.

The short-term and long-term impacts of climate change are at least of comparable importance to these notable precedents, and society will require a great amount of funding to design and administer effective mitigation programs in response to growing climate change perils. However, no effective GHG regulatory institution or economic-incentive regime intended to mitigate climate change risks can be effectively deployed without the requisite political will, which does not seem to be present at this time.

As discussed in Chapter V, a progressively increasing carbon tax could be employed in conjunction with direct regulation and public disclosure requirements to promote greater energy conservation and curtail GHG pollution, and also to help finance GHG-free replacement technologies that will eliminate many GHG discharges and enable clean-technology transfers to developing countries. We cannot successfully prevent further climate change dangers without creating substantial funding sources and improved economic incentives to deter current and future destructive activities. The point of this discussion is to emphasize the variety of ways in which pollution taxes, fees, or charges schemes could be used separately or jointly to achieve the same types of economic-incentives functions as cap-and-trade mechanisms and offset programs — to put an increasing market price on harmful GHG discharges and to create stronger incentives for businesses and consumers to replace dangerous polluting activities with more benign alternatives.

All proposed solutions for climate change hazards are bound to be complicated and controversial because of the nature of the core climate issues. No institutional approach can attain beneficial climate-policy outcomes without confronting difficult implementation and distributional problems. Carbon taxes have their own characteristic weaknesses and they must be compared against other economic-incentive strategies and direct regulation programs under realistic economic and political circumstances, not under idealized wishful-thinking conditions.

Many years ago, I coined the term "asymmetric idealism" to describe an unbalanced mode of risk assessments and social policy evaluations. Under this treatment, commentators sharply criticize competing policy choices that they consider undesirable, drawing on as much empirical evidence as available to identify the crucial shortcomings of supposedly unacceptable regulatory regimes. Yet, when the same commentators analyze and advocate their own preferred policies, they expect that everyone will cooperate when needed, that all required circumstances will come to exist, that divergent interests and priorities somehow will be harmonized, that powerful forces will not emerge to sabotage their chosen strategies or institutions, and that every desirable capability in theory will function very well in practice.

In short, asymmetric idealism assessments stress everything that has gone wrong, or could go wrong, with social policies that the commentators reject, while they assume the existence of ideal conditions and unlimited cooperation for the policies they prefer. This pattern of unbalanced comparisons cannot prove satisfactory because critical skepticism is exercised on only one side of the ledger. The other side is dominated by unjustified wishful thinking.

Every potential climate change "solution" requires careful attention to its shortcomings in comparison to alternative mitigation approaches that must perform effectively despite the contentious, short-sighted perspectives of nations, businesses, and people in a sadly imperfect world. A review of the cap-and-trade literature, for example, would show that the great majority, if not virtually all, of the proponents in favor of this market-based strategy greatly minimize the implementation and political difficulties that cap-and-trade systems must confront, while they adamantly focus on criticisms of competing regulation plans and other economic-incentive approaches.

In the endless dispute about more desirable regulatory methods, here is an illustration of "asymmetric idealism" as applied to a carbon tax. Dr. Hansen's recent book, *Storms of My Grandchildren: The Truth About the Coming Climate Catastrophe and Our Last Chance to Save Humanity,*[224] is generally inspirational and informative, but it presents a one-sided comparison between the weaknesses of cap-and-trade systems and the asserted virtues of a carbon tax scheme Hansen calls a "fee-and-dividend" system:

> In this method, a fee is collected at the mine or port of entry for each fossil fuel (coal, oil, gas), i.e., at its first sale in the country. The fee is uniform, a single number, in dollars per ton of carbon dioxide in the fuel. The public does not directly pay any fee or tax, but the price of the goods they buy increases in proportion to how much fossil fuel is used in their production. Fuels such as gasoline or heating oil, along with electricity made from coal, oil, or gas, are affected directly by the carbon fee, which is set to increase over time. The carbon fee will rise gradually so that the public will have time to adjust their lifestyle, choice of vehicle, home insulation, etc., so as to minimize their carbon footprint.
>
> Under fee-and-dividend, 100 percent of the money collected from the fossil fuel companies at the mine or well is distributed uniformly to the public. Thus those who do better than average in reducing their carbon footprint will receive more in the dividend than they will pay in the added costs of the products they buy.
>
> The fee-and-dividend approach is straightforward. It does not require a large bureaucracy. The total amount collected each month is divided equally among all legal adult residents of the country, with half shares for children, up to two children per family. This dividend is sent electronically to bank accounts, or for people without a bank account, to their debit card.[225]

Dr. Hansen presented a sample calculation finding that a $115 annual fee per ton of carbon dioxide would "increase the cost of gasoline by $1 per gallon and the average cost of electricity by around 8 cents per kilowatt-hour."[226] Yet he never discussed whether these modest cost increases could prove sufficient to persuade most consumers and industries to give up using fossil fuels in favor of unspecified alternative energy sources. These sample cost assessments fail to demonstrate that sufficient changes in consumer and commercial behaviors will occur fast enough under a fee-and-dividend program to prevent the urgent climate change hazards that Dr. Hansen has repeatedly predicted.

Our previous discussion of cap-and-trade mechanisms described Dr. Hansen's harsh but warranted criticisms of the undue influence of industry lobbyists and the willingness of many politicians to promote private interests at the expense of public needs. With regard to the fee-and-dividend system, in contrast, Hansen stated that the "backbone must be a rising fee (tax) on carbon-based fuels ... [with] no exceptions."[227] Hansen asserted that: "All sweet deals will be wiped off the books by a uniform carbon fee at the source, which will affect all fossil fuel uses."[228] Under a fee-and-dividend program that imposes a uniform carbon fee, Dr. Hansen claimed there would be "no exceptions" and "no freebies for anyone."[229] After comparing the urgency of climate change risks against the dangers "faced by Lincoln with slavery and Churchill with Nazism," Hansen claimed that "the time for compromises and appeasement is over."[230]

Dr. Hansen is certainly well-aware of the strong pressures that fossil fuel industry lobbyists and conservative pro-business ideologues have brought to bear on cap-and-trade programs in recent legislative sessions, distorting well-intentioned mitigation efforts into giveaways, windfalls, and loopholes for many of the most destructive and wealthiest corporations in the world. And yet, Hansen's description of an idealized fee-and-dividend system presumes that these powerful political influences will somehow disappear, or will be rendered comatose, as soon as a uniform tax on fossil fuels is imposed. In reality, there is no reason to doubt that socially detrimental and corrupt political practices will adversely affect Hansen's fee-and-dividend program just as they have undermined cap-and-trade initiatives. Conversely, any cap-and-trade system would surely function much better if it could be implemented in an ideal society where the forces of greed and callous indifference are taking long vacations and where politicians are all pursuing public-interest welfare.

Dr. Hansen surely is aware of the great public hostility directed at all tax increases in recent years, no matter how beneficial the purpose, and his fee-and-dividend system is an attempt to placate anti-tax politicians and voters by returning all the fee revenues to the public. However, effective mitigation programs will require substantial funding to develop and deploy "clean" replacement technologies or to force major polluters to curtail the emissions from their "dirty" technologies and processes. The Hansen fee-and-dividend mechanism does not appear powerful enough to induce sufficient emissions-reduction progress after giving all the tax revenues back to the public. His fee-and-dividend system undoubtedly will be subject to exactly the same kinds of wealth imbalances, competitive disadvantages, and political influences that were described in the discussion of cap-and-trade weaknesses. Dr. Hansen devoted considerable space to the many weaknesses of cap-and-trade approaches without offering a cogent explanation of why most of these problems would not be equally debilitating for his carbon fee (tax) strategy.

The "moral" of this comparison is that we will never achieve successful climate change mitigation until we are able to overcome the special-interest lobbyists and their supplicant politicians. We cannot simply wish them away. This combative goal obviously will not be easy to meet, but no meaningful climate progress will occur until we can successfully challenge the special-interest groups and their ideological allies, no matter which institutional mitigation regime is selected. The lobbyists and their wealthy sponsors and political minions defeated cap-and-trade programs; they have resisted direct regulation for decades; and they will just as surely attack any fee-and-dividend approach despite Dr. Hansen's optimistic one-sided predictions. Advocates of effective climate mitigation programs cannot pretend that negative political and economic forces will conveniently disappear when their favorite precautionary programs are debated, whatever those preferred programs may be.

It makes no sense for Dr. Hansen to condemn industrial lobbying and political corruption in the direct regulation and cap-and-trade contexts while he ignores the same frustrating political conditions that are equally likely to undermine his fee-and-dividend proposal. Culpable fossil fuel producers, with trillions of dollars of assets at stake and many political fellow-travelers with their hands out, will not voluntarily give up their favored positions. Nor will the political representatives of states in which fossil fuels are produced or processed, providing thousands of state jobs.[231] We cannot pretend under a fee-and-dividend scheme that political deal-makers will voluntarily forego their self-centered and duplicitous opposition to climate change programs. We have to face them and beat them, however hard that will be, to gain the broad political support essential to prevent worsening climate change dangers.

Another subject that requires a more detailed analysis than Dr. Hansen's book provided is the probable effects of the gradually increasing carbon fee imposed on fossil fuel producers or importers under his fee-and-dividend plan. Dr. Hansen has emphasized the great urgency of climate change dangers and the critical need to limit the growth of GHGs leading to potential worldwide catastrophes. He discussed the prospect of "tipping points," under which relatively small increases in the atmospheric GHG concentration may produce catastrophic non-linear effects that will be essentially irreversible.[232] He emphasized the need to phase out GHG emissions from coal combustion "as rapidly as possible or global climate disasters will be a dead certainty."[233] And he acknowledged that while the use of coal as a fuel must be eliminated immediately to the greatest extent possible, we will have to phase out oil and natural gas on a much slower timetable because we do not have sufficient clean energy alternatives.[234] Nevertheless, Hansen did not show that his fee-and-dividend plan would be any more effective or any faster in practice than direct regulation or a cap-and-trade system would be in meeting these urgent requirements under real-world conditions.

Dr. Hansen's discussion similarly did not question whether the consumer behavior modifications he predicts will arise from adoption of a uniform carbon fee would be sufficient to achieve the rapid elimination of coal-burning power plants and other coal emissions sources he contends are essential. Dr. Hansen explained how we could feasibly phase out nearly all coal usage in the US,[235] though not in the large GHG-polluting developing countries, but he did not show that the slowly escalating carbon fee in his fee-and-dividend program would impose high-enough fossil fuel energy costs and would create strong-enough consumer incentives to achieve this necessary goal.[236] In the same vein, Hansen stated that "most of the fossil fuels must be left in the ground"[237] without demonstrating that his fee-and-dividend system would create strong-enough deterrent incentives over a short-enough time-horizon to accomplish this objective.

Dr. Hansen apparently has become so critical of government and business that he now places the main responsibility for reducing GHG emissions on consumer behavioral changes he claims will be induced by the increasing annual carbon fee. In one passage, he observed that the "beauty of the fee-and-dividend approach is that the carbon fee helps any carbon-free energy source, but it does not specify these sources; it lets the consumer choose."[238] He contended that "in the long run" people "will tend to adjust their decisions on vehicle choice and other matters as the carbon price gradually continues to rise."[239] Dr. Hansen similarly claimed that a carbon fee "internalizes" the "incentive to reduce the use of carbon fuels, especially coal, in billions of energy decisions ranging from commuting behavior to the design of vehicles, aircraft, cities, and so forth."[240]

There is a large cognitive psychology literature that indicates most consumers will not have the time, attention span, or expertise required to make effective choices in as complicated a context as global climate change.[241] Even if typical consumers make "good" decisions 50% of the time with regard to climate change choices, which is a highly optimistic supposition, they would be making "bad" decisions as often as "good" ones. It is difficult to conceive how innumerable consumers could make "literally billions of decisions," as Dr. Hansen claimed, in an informed manner despite the complex and uncertain characteristics of almost all climate change issues.

With regard to the dividend part of the plan, Dr. Hansen contended that: "Returning the money to the public is the hard part in the United States. Congress prefers to keep the money for itself and divvy it out to special interests."[242] Only Congress has the power to adopt the fee-and-dividend system into US regulatory law, and the legislators can refuse to make this controversial climate-policy decision just as they have refused to accept a cap-and-trade system.

Affected special-interest groups undoubtedly would do everything they can to defeat or warp Hansen's fee-and-dividend approach, which he claims could substantially impact their economic and political welfare. Unfortunately, it is wholly implausible to expect that Congress would enact a politics-free, public-interest regarding, uniform carbon fee, and would then hobble its "power of the purse" by giving all the fee revenues back to US citizens on an equal basis.

In one of his most unrealistic paragraphs, Dr. Hansen claimed that under a gradually increasing fee-and-dividend system, "we will reach a series of points at which various carbon-free energies and carbon-saving technologies are cheaper than fossil fuels plus their fee. As time goes on, fossil fuel use will collapse, remaining coal supplies will be left in the ground, and we will have arrived at a clean energy future."[243] In addition to this unbelievable vision, Dr. Hansen maintained that the "best enforcement is carbon price — as the fuel price rises, people pay attention to waste."[244]

If only overcoming the widespread use of fossil fuels were this simple, which it is not. There has been an extensive debate in the literature on climate change about the relative effectiveness of carbon tax schemes in comparison to cap-and-trade systems,[245] which includes some of the observations presented in this chapter. However, remarkably little discussion has focused on how the existing wealth and political influence of fossil fuel producers would affect the performance of any kind of economic incentive program based on market-system processes.

The fossil fuel producers are among the wealthiest and most influential companies in the world; indeed, several US Senators recently complained that the large oil companies have made a trillion dollars in profits over the past decade.[246] Yet, Dr. Hansen failed to explain why these business giants would not use their economic strength and other market advantages to undermine less well-established

GHG-free energy companies by keeping fossil fuel prices relatively low despite the increasing carbon fee. If renewable energy companies and other GHG-free enterprises are unable to compete with the fossil fuel companies, or are unable to attract sufficient investments to sustain this competition for many years, consumers often will not have any realistic alternative energy sources to which they can switch. Consumers will not be able to discard fossil fuel energy, as Dr. Hansen blithely imagines, because the American people must have access to sufficient energy and they may not have any other realistic way to obtain this energy under a market-forces regime influenced by, if not dominated by, powerful fossil fuel industries except by continuing their reliance on fossil fuel technologies and practices.

This concern for the effects of unequal market forces is more than conjectural. In the month of August 2011, for example, three major American manufacturers of solar energy panels were forced to declare bankruptcy because of inadequate market demand in the US.[247] If GHG-free clean energy companies cannot remain financially viable in the face of unrelenting market competition from much wealthier fossil fuel producers, the "dirty" companies will be able to raise their prices to cover Dr. Hansen's escalating carbon fees and will then pass on their higher operating costs to consumers who lack genuine choices. The same consideration applies to cap-and-trade systems, in which a substantial portion of the carbon allowances or offsets could be purchased by fossil fuel companies and their allies. Dr. Hansen does not explain why his proposed fee would transform weak renewable energy companies into market tigers.

Dr. Hansen also neglected to address the fact that many of the heaviest users of fossil fuels, especially coal, are regulated by government monopolies in the form of public utility commissions. The electric energy from coal-burning power plants is normally regulated by state utility commissions that can authorize energy price increases on a cost-plus basis, which means energy prices would increase commensurately when coal or oil prices rise as a result of the hypothetical increasing carbon fee or cap-and-trade allowances price. In most states, consumers have little choice about which energy sources they must use, and the choices they do make have little influence on the energy regulation process. There is no reason to expect that most state utility commissions will choose to ignore the ties they have established with fossil fuel producers over many decades.

The fossil fuel producers and GHG-free renewable energy companies would not be competing on anything resembling a so-called "level playing field." The fossil fuel companies possess enormous wealth, well-established product distribution networks, economies-of-scale, strong political leverage, many millions of current customers, and other important economic advantages that will probably enable them to prevail indefinitely under competitive market forces unless the carbon taxes or tradable allowances prices are set at such a high level that they really ought to be described as direct regulation mandates rather than economic incentives options. The talismanic phrase, "putting a price on carbon," has been vastly overhyped and will only be useful when the carbon tax or market price of carbon allowances is high enough to provide GHG-free energy sources with a substantial competitive advantage that would outweigh all of the existing market advantages enjoyed by fossil fuel producers and sellers.

Dr. Hansen devoted no serious attention to the prerequisite conditions for the success of his fee-and-dividend plan, and the same lack of careful economic analysis applies to many positive assessments of cap-and-trade systems and other economic incentives measures. Hansen's wishful-thinking description of how increasing the carbon fees will raise the prices of fossil fuels and eventually will create the market conditions necessary for their replacement by GHG-free energy technologies implicitly presumes that the wealthy fossil fuel companies would be sitting on their hands and passively acquiescing while this gradual transformation takes place. Sadly, that is not a believable scenario.

Notwithstanding the arguments presented here against Dr. Hansen's fee-and-dividend proposal, I similarly prefer a carbon tax approach to a cap-and-trade system. On the other hand, I do not support his "dividend" program, which fits into the general category of rebate or feebate programs meant to mollify tax-paying consumers.[248] Giving a rebate to poor people, who otherwise would be seriously hurt by a regressive carbon tax, could be worthwhile and perhaps politically necessary but I strongly disagree with the idea of returning all of the fee revenues to American citizens based on a simple formula. Dr. Hansen contended that the 100%-taxpayer-return dividend can be a comprehensive solution for US and international discharges; and yet, climate change regulation will require a substantial portion of the tax money to fund the creation of clean GHG-free technologies and technology transfers that will enable poor countries to stop following the historical pattern of "dirty" development and climate degradation. Where else or how else are climate mitigation programs going to obtain the financial support required to achieve these essential functions?

Every proponent of carbon taxes in preference to cap-and-trade systems argues that tax instruments would be less complicated and much easier to understand and adjust depending on feedback from the experiences of the first few years. However, the most crucial difference is that under a cap-and-trade scheme, unlike under a tax system, thousands of participating actors will be engaged in pursuing their own self-interest and profit-making opportunities as much as or more often than in attaining climate mitigation improvements.

Dr. Hansen rightly criticized the Wall Street moguls for threatening to turn tradable GHG allowances into financial derivatives and other fiscal gimmicks whose only purpose is to make profits for speculators. But he did not mention the other categories of participants who may also profit from this economic incentives approach, including the people who set up and maintain the carbon allowances and offsets markets, the people who develop the emissions-reduction projects and try to obtain approval for the allowances or offsets from these projects in various developing nations, and the people who are supposed to monitor and enforce allowances trading or offset implementation to ensure that the sponsors and beneficiaries of these programs are really getting what they are paying for. It is, for example, no exaggeration to say that many CDM consulting companies have been swarming through dozens of developing nations to assist the recipient states in qualifying for CDM projects and payments, while the consultants earn significant commissions and the investor nations claim that they are "offsetting" their GHG discharges.

Any system where there are so many opportunities for maximizing personal profits at the expense of environmental goals should be evaluated with great critical skepticism. Even sincere and altruistic agents may begin to look for personal advantages after a few years of frustration caused by constant exposures to political corruption and short-sighted greed in participating nations, including the US. My experiences in the field of international biodiversity conservation suggest that there is a high "burn-out rate" among the people dedicated to "doing good," and I see no reason to expect that the performance and incentives of the carbon market participants focusing on climate change programs will be any different.

The people who are supposed to monitor and validate this process can make a respectable salary by approving whatever projects or applications they find, but they would be jeopardizing their careers if they reported violations that threatened the viability of any carbon allowances or offset projects. The pressures "to go along to get along" would certainly be strong in the case of CDM projects and other allowances and offsets initiatives. Everybody in the field would win by facilitating or approving CDM initiatives and other offsets and allowances, whether justified or not. Everybody would win except the climate change victims. In contrast, a carbon tax or fee program would typically be administered by a government agency and the incentives of the participating agency officials would be markedly different from those of the private people participating in market-based cap-and-trade or carbon offset systems.

It is fair to conclude that the flexibility and profit-motivation afforded to GHG dischargers under a cap-and-trade system or carbon offset program would make these ostensibly "efficient" programs the most complicated, costly, easily obstructed, and difficult-to-maintain emissions-reduction approaches from the perspective of implementation, monitoring, and enforcement requirements. These operational problems reinforce an important conclusion I published about 25 years ago: The most efficient regulatory strategy *in theory* is often the least effective *in practice* because it requires more information, cooperation, administrative attention, funding, compliance monitoring, and enforcement efforts; and also because it is usually more vulnerable to intentional delays and obstructive behavior by regulated parties.[249] All of these weaknesses apply to economic incentive initiatives in the climate change context.

With all of the factors strengthening the competitive positions of fossil fuel producers and their long-standing users, it is difficult to see why many people remain convinced that "putting a price on carbon" would tip the critical market-balance in favor of non-destructive clean energy sources. For example, this doubtful presumption underlies Dr. Hansen's fee-and-dividend proposal, in which he blithely asserted that the fee will gradually drive the fossil fuel companies out of business and will result in leaving almost all US coal in the ground. We need to understand in detail whether, and why or why not, a relatively modest carbon fee could revolutionize the economic and energy structures of the US and the world, as Dr. Hansen claimed, before we commit ourselves to following any market-based mitigation approach.

I want to emphasize my respect and appreciation for Dr. Hansen's scientific knowledge, personal courage, and unwavering commitment to overcoming climate change risks. He has persevered for decades despite confronting many indifferent politicians and corrupt special-interest lobbyists, and he deserves great credit for raising the public's awareness of climate change issues. However, in my opinion, his recommended "fee-and-dividend" solution is wholly unrealistic. An idealized solution that ignores almost all of the difficulties that have undermined previous mitigation programs is no solution at all. Instead of relying on wishful thinking, we must examine the operation of economic incentive mechanisms under realistic market conditions in which the fossil fuel companies have much greater economic strength than the producers of GHG-free technologies and the advocates of climate change mitigation.

Allowing the fossil fuel producers to win a market-based energy competition would in effect authorize the ongoing degradation of climate change conditions, with incalculable human and environmental harms. This would be intolerable. Our society cannot afford to rely primarily on market system choices to ensure desirable climate outcomes. Instead, we must impose strict regulations on major GHG-polluting sources, and we could combine this mandate with carbon taxes or tradable allowances to create additional deterrent incentives and to raise the funding needed for the development of GHG-free replacement technologies and processes.

This is another way of saying that no economic incentive mechanism, whether taxes, fees, offsets, cap-and-trade allowances, or any combination of them, would be sufficient to protect against increasing climate change risks as long as we allow fossil fuel producers and their allies to exploit their wealth and political advantages in the market system. In a competition between fossil fuel producers and GHG-free alternative technology producers, we cannot allow the market system to decide which side will prevail regardless of the climate change dangers. Otherwise, there is a strong likelihood that fossil fuel companies and their political and economic supporters will be able to continue their business operations regardless of the long-term climate degradation and human harms they have been causing for decades.

As explained in greater detail in Chapter V, the most plausible way to prevent this wealth imbalance from dominating market forces and insulating the fossil fuel producers from the need to adopt clean energy methods would be to impose direct regulations on the major sources of GHG emissions together with carbon taxes on large and mid-sized polluters. The regulations must be designed to overcome the weaknesses in "free markets"; the economic incentive system must be designed to overcome the weaknesses in direct regulation; and these mitigation institutions must be designed to function together. This climate-policy prescription will certainly not be easy or inexpensive to attain, but Chapter V argues that a multi-institutional structure with overlapping responsibilities represents a more realistic strategy than either market-based economic incentives programs or direct regulation working alone in this difficult public-policy domain.

One advantage of a cap-and-trade system over a carbon tax is that the cap sets a distinct limit on the cumulative amount of GHG discharges that are authorized each year, while the carbon tax would allow as much GHG pollution as producers and sellers choose to discharge as long as they can pay the applicable tax for their desired emissions volume. Let me note that the cap is entirely a regulatory function, not a market mechanism, in which maximum GHG limits are set by government decree and the government decides whether to increase or decrease the number of GHG allowances as experience is acquired. Only the trading of emissions allowances above the designated cap would entail the use of economic incentives and market forces. Above-the-cap discharges authorized by tradable GHG allowances, or discharges authorized by payment of GHG taxes, are equivalent to the persistent residual emissions that will cause the atmospheric GHG concentration to grow progressively higher. If the number of allowances is reduced every few years, or if the tax level is increased, which would be another regulatory function rather than a market achievement, this regulatory pattern might "reduce the increases" in GHG discharges in the air but it would not reverse the climate degradation process.

I believe the practical distinctions between cap-and-trade systems and carbon taxes are not especially compelling or likely to yield appreciably different climate benefits. However, it is important to note that GHG allowances trading programs have not performed well in Europe,[250] where they have been the core of the EU Kyoto Protocol and Emissions Trading Scheme (ETS) compliance strategies for years.[251] Despite this spotty history,[252] I have not found any American cap-and-trade advocate explaining in detail why our proposed GHG cap-and-trade programs would be implemented more effectively or would perform any better than the cap-and-trade experiences of the EU. As a consequence of the respective political cultures, and especially the contentious character of American politics, it is likely that antagonistic political opposition and interest-group pressures will be even stronger in the US than they have been in Europe.

Every climate-oriented mitigation program must be able to reduce the atmospheric GHG concentration or it cannot overcome growing climate change dangers. Partial reliance on economic incentives measures could increase pollution deterrence by imposing higher prices on GHG dischargers and it could raise funding to support transformative GHG-free replacement technology programs. In this regard, a relatively simple carbon tax could probably be integrated into a multi-institutional regulatory plan more easily than a cap-and-trade system could be, and we would not have to worry as much about personal profit-oriented incentives created by complicated combinations of GHG mitigation efforts and self-enrichment motivations.[253]

CHAPTER IV

The Stalemate in International Negotiations

Since the UN Framework Convention on Climate Change was adopted in 1992,[254] nearly constant negotiations have been aimed at creating a generally acceptable international policy to counter global climate change. Highly-publicized summits attended by many heads-of-state of UNFCCC signatory nations have been held in Kyoto, Bali, Copenhagen, and Cancun; and secondary meetings on climate change issues were held in Barcelona, Berlin, Bonn, The Hague, Marrakesh, Oslo, Poznan, Seoul, Tianjin, and other locations. These climate-policy negotiations have featured wide-ranging discussions about mitigation, adaptation, funding, monitoring and verification, technology transfers, and institutional or programmatic designs among thousands of diplomats with conflicting national interests and priorities. Aside from the Kyoto Protocol,[255] which established a first phase of minimal emissions-reduction obligations for developed states that are expected to expire in 2012, and the voluntary Copenhagen Accord,[256] which asks all nations to commit to their own chosen levels of GHG pollution control, little international agreement has been achieved because of crucial differences in the positions of the major blocs of nations. Nevertheless, the practice of holding an annual succession of international meetings on climate change problems with a plethora of interstitial negotiations is still flourishing.

There is widespread agreement that climate-policy disputes between developed nations and developing nations, often called "Northern" and "Southern" states, will be the most difficult impasse to bridge through negotiations. These blocs are also described as the

38 developed Annex I nations under the Kyoto Protocol and the developing non-Annex I countries. The Kyoto Protocol distinction has become quite important because many developing nations, especially China and India, are determined to retain their current status as non-Annex I states with no international emissions-reduction obligations regardless of their rapidly growing economies and commensurately increasing GHG discharges.

The national positions within these blocs are by no means monolithic, with many of the poorest developing nations that are most vulnerable to climate change dangers objecting to the intransigent positions advocated by the large GHG-polluting developing countries of Brazil, South Africa, India, and China, collectively referred to by the acronym as the "BASIC" states. In another divergent position among developing nations, Saudi Arabia has been recruiting oil-producing states to support the claim that they should receive compensation if oil usage is reduced as a result of climate change mitigation, and "[t]hey view any attempt to reduce carbon dioxide emissions by developed countries as a menace to their economies."[257] Despite policy variations in some specific contexts, the long-standing economic inequalities and political disputes between the affluent developed nations of the North and the relatively less-developed nations of the South continue to provide the core foundation for international relations disagreements in nearly every substantive field, including global climate change.

In a commentary summarizing the results of the 2009 Copenhagen meeting, Professor Daniel Bodansky, a widely-known American international law scholar, emphasized the recurring disputes between the Southern and Northern blocs of nations:

> On one side, developed countries insist that the post-2012 regime address the emissions of all of the major economies, developing as well as developed. On the other side, developing countries continue to argue, as they have done since the negotiations first began back in 1991, that they are not historically responsible for the climate change problem, have less capacity to respond, and hence should not be expected to undertake ... international emissions reduction commitments.[258]

Less dispassionate commentators have expressed disappointment and anger at the inconclusive Copenhagen results, shifting their characterizations from "Hopenhagen" to "Brokenhagen," "Tokenhagen,"[259] and "Nopenhagen," while criticizing many participating nations that "will predictably choose the view that suits their wider agenda."[260] The following assessment primarily supports Northern bloc goals while attacking the positions of China and other influential Southern states at Copenhagen:

> It became increasingly clear, when following the discussions and arguments advanced by the various delegations, that national interest superseded the global obligations every nation has to look after our common planet, and calling demands for accountability and responsible transparency "meddling in internal affairs" or suggesting a "loss of sovereignty" is enough of a giveaway of their intransigent and stubborn stone walling, which already emerged at the recent summit of the Pacific Rim countries in Singapore.[261]

In contrast, a representative of the government of Tuvalu expressed dismay at the continuing unwillingness of the United States to make tangible commitments at Copenhagen to reduce its GHG emissions rapidly despite the harm those discharges are doing to small developing nations that cannot solve climate change problems on their own. Tuvalu and more than 100 other developing countries do not discharge a large-enough volume of GHGs to make any significant difference in worldwide emissions-reduction mitigation efforts, and yet they are among the most vulnerable victims of climate change harms and the most frustrated critics of the lack of movement toward meaningful climate change progress. As one critic of the Copenhagen Conference contended:

> It was very evident that President Obama had nothing to bring to the negotiating table. Without the US on board to a legally binding outcome, it was very evident that large emitting countries were not going to stand up and agree to take significant emission reductions commitments or actions.
>
> We must move beyond the US impasse. Diplomatic efforts must be applied to the Obama Administration to encourage it to assert its

> executive power and commit to emissions reduction commitments based on international standards. Otherwise the rest of the world has no option but to isolate the US. Applying trade-related measures to discriminate against embedded emissions in US products could be one option.[262]

I can feel sympathy for this David versus Goliath challenge, but there is nothing that the Small Island Developing States (SIDS) can do in practice to impose more responsive climate-policy obligations on the US government. Threats intended to force a recalcitrant nation to comply with other states' environmental goals and standards are usually incompatible with basic international law principles, including national sovereignty and autonomy, as the US discovered when it tried to protect endangered sea turtles and tuna stocks against depredations by short-sighted fishing countries. A central premise of international law is that multilateral commitments must be based on voluntary agreements, and states cannot be bound to legal obligations that they have not willingly chosen to undertake.

An illustration of this principle of broad sovereignty is reflected in the climate-policy priorities of the national government of India, a non-Annex I developing state with a rapidly growing economy and commensurate GHG emissions increases. Here is one interpretation of India's response to the Copenhagen Conference outcome from a well-informed Indian climate-policy analyst I met in New Delhi:

> Minister of State for Environment and Forests, Jairam Ramesh, has offered the conclusion that the Copenhagen Accord is a "good deal" for India.
>
> This perspective seems to be based on the claim that India's "redlines" of no [GHG emissions] cuts, no nationally specific [emissions] peaking year, and no international verification had been met. And indeed, from the early news reports and from a debate in the Rajya Sabha [the upper house of India's Parliament] on the Copenhagen meeting, it is on these redlines, and on the question of international verification, in particular, that the national debate is engaged.[263]

In other words, the Indian climate-policy negotiations position was based on the rejection of specific GHG emissions-reduction commitments, specific future limitations on the growth of GHG emissions, and any form of international verification applied to India's emissions levels or other commitments the Indian government may choose to make with regard to any climate change initiative.

I visited India for six weeks in February and March of 2010 and discussed climate change issues with many Indian scientists — I was able to attend the first national conference of Indian climate change scientists, held in New Delhi in early March — as well as policy analysts, government officials, business people, and some Indian working people chosen randomly. I found that many Indian officials and citizens are well-aware of the severe dangers that climate change creates for their country and of the need for better mitigation and adaptation programs in India. However, to an overwhelming degree, government programs placed greater emphasis on climate-policy words than on mitigation or adaptation actions, and I saw no indication that the government is considering shifting substantial resources from planned investments in more economic development to any significant investments in climate-related precautionary measures.

This experience reinforces an important conclusion: The central disputes in international climate-policy negotiations to a large extent reflect the deeper conflicts at the heart of all North–South relations. Almost all developing states are committed to increasing their economic and social welfare to promote greater prosperity as well as poverty alleviation, and they are subordinating their climate change positions or protections to these general economic goals.

Nearly all developing nations are also adamant about maintaining national sovereignty and autonomy, which is hardly surprising because many of these states were once victimized by colonialism and imperialism, or feel that they were. Their positions on vital climate issues, such as how international funding for mitigation and adaptation programs should be controlled, show their strong desire to avoid intrusive "strings" imposed by former colonial or hegemonic powers, which for the most part are the affluent developed nations of the North.

Nearly all developing states stress the need for distributional equity and fairness by asserting that the North has been, and still is, consuming an unfair proportion of the Earth's limited resources. These complaints are repeatedly raised by developing nations in negotiations aimed at attaining more foreign assistance and investments, better trade terms, flexible intellectual property rights, technology transfers to facilitate greater development, and other disputed issues that the less-affluent nations consider *as important to them* as climate change policies.

On the other side of the ledger, the countries of the North want to achieve climate change improvements in ways that do not seriously threaten their economies, employment, and standards of living. They want to avoid creating competitive disadvantages resulting from expensive restrictions on their own polluting industries, while the foreign competitors in developing nations need not meet any GHG emissions-reduction requirements. Many developed nations are willing to give substantial mitigation and adaptation funding assistance to cooperative developing countries. However, the North wants to ensure that their foreign aid will actually be used effectively for the purposes intended. This is a serious concern because well-documented case-histories show widespread public and private corruption in many of the developing states that would receive assistance for climate programs. Representatives of the North consequently make frequent demands for good governance, accountability, transparency, and better MRV capacities as preconditions for financial and technological benefits agreements, while the Southern states regard these intrusive conditions or "strings" as interferences with their national sovereignty and internal government autonomy.[264]

There is no reason to believe the North–South stalemate among countries with divergent perceived national interests has been, or will be, resolved by more negotiations to any significant extent,[265] and the lack of a consensus enabling strong collective international actions will prove harmful for all nations and peoples. For decades, diplomats representing the Southern and Northern blocs have been constructing new arguments and strengthening existing arguments to support their positions on conflicting international relations issues including diverse climate mitigation treatments. No magic wand will make the international blocs modify their priorities as long as they continue to pursue the customary self-interested national goals. I question whether piling more arguments on one side or the other will lead to successful negotiations in the absence of fundamental changes in perceived national interests among the major participants.

Before we examine the specific arguments presented by each bloc in greater detail, it may be useful to introduce an original perspective that I call the "political suicide threshold." If the leader of a developed country unilaterally attempts to reduce the wealth and natural resources access of that state substantially, in order to meet the desires of less affluent nations for greater economic development, would this leader really be helping to achieve a worldwide redistribution of economic resources and benefits or would the leader more likely be committing political suicide because of the opposition of the country's population and other policymakers to accepting the necessary radical sacrifices?

This is not merely a hypothetical question. Some climate-policy analysts in developing nations argue that the North must sharply reduce its GHG emissions on a per capita basis to reach a fair and equal allocation of the limited atmospheric capacity to absorb greenhouse gases. Because developing states have several billion more people than the affluent nations of the North, this per capita emissions-reduction policy would greatly curtail the amount of permissible GHG emissions each developed nation could discharge, which would have severe debilitating effects on the wealthy countries' economic productivity and living standards.

Other developing nations now argue that because the North is much wealthier than the South, the developed nations must cut their GHG emissions by even more, much more, than under a per capita allocation policy. Instead, these developing states contend that the North must cut its GHG emissions enough to leave a substantial amount of "carbon space" or "development space"[266] that would allow relatively poor nations to increase their GHG pollution *above a per capita target* as a way to attain greater economic growth. Under this "carbon space" treatment, the countries of the North would have to reduce their emissions appreciably below the worldwide per capita average, which would be economically disadvantageous, if not disastrous, for their economies and employment prospects. My question is whether this is a realistic expectation to impose on the governments and people of affluent countries, or whether it would be much more likely to lead to the political suicide of any national leader in the North who advocates this radical redistributive climate-policy position.

Now let us imagine instead that the leader of a large developing nation, with steadily increasing GHG pollution resulting from rapidly expanding economic growth, chooses to set a specific stringent target for GHG emissions reduction, as the Northern countries have repeatedly proposed, at the cost of slowing or eliminating further national economic development. My discussions in India reinforced the view that the leaders of developing nations, especially democratic ones, would be committing political suicide by acceding to the demands of the North at the expense of undermining the most important economic growth goal of nearly all Southern states.

When evaluating the self-serving but not necessarily invalid arguments presented by the Southern and Northern nations, readers should consider whether any international agreement on substantial climate change mitigation programs is possible that would not require the responsible leaders on one side, or perhaps both sides, to bypass the realistic "political suicide threshold." The South and North bloc climate-policy divisions may arguably be so deep that fundamentally conflicting positions can never be harmonized by following the current patterns and positions of international negotiations.

Many national leaders may have much less bargaining flexibility than people on "the other side" imagine because most politicians are unwilling to jeopardize their official positions. Consistent with the political suicide threshold, no important climate-policy agreement may be attainable as long as the major nations of both blocs continue pursuing their narrowly-defined national interests and espousing the same types of climate-control measures that have previously been rejected countless times by the opposing blocs.[267] As the Copenhagen Conference demonstrated, changing the words employed in successive negotiations will seldom improve the unsatisfactory results as long as the perceived national interests and proposed climate mitigation strategies remain essentially the same.

Arguments Supporting the South's Positions

These arguments have been used by the South bloc to justify their unwillingness to make specific emissions-reduction commitments and their expectations to receive substantial financial assistance from developed nations that are largely responsible for the increasing growth of GHG pollution and climate change dangers:

Historical Responsibility: In 1992, the UNFCCC treaty text included an introductory provision that emphasizes historical emissions:

> *The Parties to this Convention ... Noting* that the largest share of historical and current emissions of greenhouse gases has originated in developed countries, that per capita emissions in developing countries are still relatively low and that the share of global emissions originating in developing countries will grow to meet their social and development needs ...[268]

It would be difficult to refute the argument of the South that the developed nations have created the climate change problem as a result of their long-term industrial growth and high-consumption lifestyles.[269] For example, one report contended China's position is that "the West is responsible for rising temperatures, because it has been pumping climate-changing gases into the air for centuries."[270] In the same vein, the 2008 Indian National Action Plan on Climate Change (NAPCC)[271] attributed "the global threat of climate change" to "intensive industrial growth and high consumption lifestyles in developed countries."[272] A prominent Indian policy-analyst subsequently observed: "there is a structural tension around whether and how, the global climate regime should reflect the fact that different countries carry different levels of responsibility for causing the problem."[273] If the affluent developed countries are primarily to blame for climate change dangers, why should the poor developing states of the South have to provide the solution?[274]

Increased Economic Development and Poverty Alleviation: Increasing the economic and social development of relatively poor nations has become the UN's highest priority,[275] with the possible exception of peace-keeping efforts, and the right to greater development has been incorporated into numerous international treaties. This is the foremost goal of almost all Southern nations. In June 2010, for example, UN Secretary-General Ban Ki-moon urged the "leaders of the biggest industrialized and developing nations to focus on development, green growth and the needs of the most vulnerable, in devising [economic] recovery strategies"[276] and only later in this message he mentioned that such "an approach can help address food security and climate change, while ensuring job creation."[277] I think this is a fair reflection of the hierarchy of priorities embraced by the Southern bloc of nations.

Dr. Rajendra K. Pachauri, the leader of the IPCC who accepted their half of the 2007 Nobel Peace Prize, has argued that: "Developing nations must be allowed to boost carbon emissions to lift millions out of poverty."[278] Jeffrey Sachs, a Nobel Prize-winning economist, similarly contended: "If we try to restrain emissions without a fundamentally new set of technologies, we will end up stifling economic growth, including the development prospects for billions of people."[279] Regardless of whatever normative principles people may embrace, the current "real world" position is that most GHG-polluting developing nations will not accept stifled development prospects resulting from the North's climate policies.[280]

Rather than accepting GHG emissions-reduction commitments that could restrain further economic growth, as proposed by the North bloc, the nations of the South contend that wealthy developed countries must take the lead in reducing GHGs. The leading Chinese negotiator at the Copenhagen Conference contended that "China's national interests will always come first and, in any move toward binding steps for reducing global emissions of greenhouse gases, rich countries must go first."[281]

The Southern nations also argue that the North should pay for whatever climate mitigation or adaptation efforts the developing states may choose to undertake as well as paying for their own climate change programs.[282] In response to poverty issues, the South maintains that it would be preferable to reduce GHG emissions from nations with prosperous high-consumption standards of living than from comparatively poor countries with many millions of people still struggling on the edge between destitution and subsistence survival.[283]

Equity and Fairness: If the amount of GHGs the atmosphere can absorb is inherently limited, why should some wealthy nations be able to exploit a much larger proportion of the capacity on a per capita basis than other states? For example, India generates only a little more than one-twentieth of the current per capita GHG discharges from the US and Australia, and India with hundreds of millions of poor people has contributed even less on a per capita basis to the atmospheric build-up of greenhouse gases during the past century.[284] It would be more equitable if countries with disproportionately large per capita discharges had to pay the nations with low per capita discharges in return for exceeding their fair share of the aggregate atmospheric GHG capacity. One Indian newsletter observed: "From the 1990s there has been a consensus that freezing emissions at current levels would mean freezing inequity. Climate justice demanded that the rich reduced emissions, so that poorer, emerging world could grow."[285]

Reparations and Damages: Because the developed nations have produced the great majority of the GHG emissions that are causing global climate change and are continuing to discharge higher pollution amounts on a per capita basis, they should have to pay the South for the severe climate change harms already inflicted on many developing countries and for the future climate risks they have imposed. This treatment would be consistent with the "polluter pays" principle that has been widely accepted in public international law in theory though rarely in practice.[286]

Common but Differentiated Responsibilities: The UNFCCC incorporated the proviso that the developing countries cannot be expected to achieve the same climate change mitigation measures as Annex I developed nations:

> *The Parties to this Convention ... Acknowledging* that the global nature of climate change calls for the widest possible cooperation by all countries and their participation in an effective and appropriate international response, in accordance with their common but differentiated responsibilities and respective capabilities and their social and economic conditions ...[287]

The Kyoto Protocol reaffirmed the "common but differentiated responsibilities" (CDR) principle in Article 10 and added that states-parties should "take into account" their "specific national and regional development priorities."[288] The Kyoto Protocol also limited the obligation for "quantified emission limitation and reduction commitments" only to the developed nations identified in Annex I.[289]

In essence, the Southern bloc contends that the formative international agreements on global climate policy 20 years ago explicitly limited quantitative emissions-reduction obligations to the developed nations and allowed developing nations to choose whatever mitigation, adaptation, and funding responsibilities individual non-Annex I countries may prefer in light of their other national objectives and priorities.

The Copenhagen Accord, in contrast, did not reaffirm the CDR principle and instead asked all nations from the North and South to make voluntary emissions-reduction commitments. This controversial negotiated treatment may partly explain why the Accord was not approved by most participating states at the Copenhagen meeting, and why the last-minute voluntary agreement was accompanied by the inconclusive language that the member-states "took notice" of the Accord rather than formally adopting it.[290]

Constructive Efforts to Address Climate Change Problems: Most of the largest GHG-polluting developing nations claim that they are already taking meaningful steps, consistent with their capacities and priorities, to reduce their contributions to climate change dangers. For example, although the Copenhagen Accord was voluntary, China made a commitment to increase its "carbon intensity" by around 40–45%, and India followed suit by promising roughly a 20–25% intensity improvement. Some experts regarded these promises as significant steps toward greater economic and energy efficiency with lower GHG emissions,[291] while other commentators viewed the self-imposed commitments less favorably.[292]

In evaluating the Copenhagen meeting, Roy Lee, the former Secretary for the UN Conference on the Establishment of the International Criminal Court, and former Secretary of the UN General Assembly Legal Committee and International Law Commission, argued: "All the world major emitters, including the 38 Annex I KP [Kyoto Protocol] parties, have signed on to the Copenhagen Accord and are all promised to reduce emission either on a fixed rate basis (e.g., the EU), from business-as-usual (e.g., Brazil and Korea) or in carbon intensity (China and India). Altogether, over 114 countries have formally associated with the Accord."[293]

Improving carbon intensity does not ensure that GHG emissions will be reduced or that the atmospheric GHG concentration will be lowered. Intensity represents the degree of energy efficiency or economic productivity associated with a given amount of GHG discharges. If more energy or economic output is produced for each unit of GHG emissions, that does constitute improved efficiency because more benefits will be gained from each given volume of GHGs.[294] However, if energy or Gross Domestic Product (GDP) output expands by 10% a year, as China and India are planning to achieve, and if the intensity is hypothetically increased by 50% (greater intensity than they have pledged), these countries would still be putting out 5% more GHG emissions every year in addition to their annual BAU discharges, and consequently they will be making climate change worse. In effect, these nations would be discharging the other residual 50% of BAU discharges resulting from their increased productivity. The higher carbon intensity is *better than nothing* compared to putting out 10% more GHGs annually to match the expected economic growth rate, but the improvement in carbon intensity would nonetheless frequently produce a "reducing the increases" impact on the atmospheric GHG concentration and would not reduce any climate change hazards. As long as the economic output in a developing country continues to expand, even large intensity improvements may not reduce the cumulative volume of annual GHG emissions or that nation's contribution to the atmospheric GHG concentration.

In two other large developing states, Brazil and Indonesia, most GHG emissions come from deforestation and unsustainable land-use practices. These nations have pledged to cut the deforestation rate sharply in return for substantial funding from international sources, but both have long histories of widespread illegal logging, government indifference to environmental conditions, and frequent corruption among business and government elites. There have been nearly constant negotiations for the past few years about the implementation of effective MRV programs to ensure compliance with national deforestation or reforestation commitments, but the markets for timber will continue to exist in Brazil and Indonesia while the asserted GHG emissions-reduction benefits of anti-deforestation projects could easily be undermined by legally or illegally felling more trees somewhere else in the vast forests of these countries.

The Northern countries have placed substantial emphasis on the need for MRV measures and better governance as prerequisites for major international funding programs to improve GHG emissions reduction in developing states, or to combine economic development assistance with climate change mitigation programs. However, the Northern nations have been reluctant to question publicly whether most Southern states could implement reliable climate-control measures even if they receive ample funds and want to meet the North's objectives. There certainly has been no resolution of the central reality that the South wants much greater development assistance on redistributional and equitable grounds without bearing a plethora of financial "strings," while some of the large GHG-polluting developing states are not very interested in addressing climate change problems that they blame overwhelmingly on the Northern countries.

Arguments Supporting the North's Positions

Historical Responsibility: The Northern countries have argued that climate change dangers were not widely recognized until the 1970s at the earliest, and the comprehensive dimensions of the climate problems did not begin to emerge until two decades later. As one example, the lead Obama Administration negotiator, Todd D. Stern, contended that: "For most of the 200 years since the Industrial Revolution, people were blissfully ignorant of the fact that emissions caused a greenhouse effect. It's a relatively recent phenomenon."[295] In other words, anthropocentric climate change only emerged as a serious world threat in the past few decades and the developed nations should not be blamed for unknowingly causing it.

Though few climate experts in the North would dispute their bloc's primary causation of climate change problems, some have argued that the South has also made major contributions to increasing climate risks by allowing widespread deforestation, destructive agricultural practices, extensive livestock grazing, and other forms of harmful natural resources exploitation. In the same vein, one senior staff member of the Brookings Institution raised the problem of India's explosive population growth (from 350 million people in 1950 to over a billion by 2000), and asked: "If developed nations are held responsible for emissions that they historically contributed, oblivious to their impact on climate change, why shouldn't developing nations take responsibility for producing generations of people who will generate more GHG emissions into the future?"[296] This comment implicitly stresses the fact that the overwhelming majority of the world's population growth is taking place in the South, and the growing population is bound to demand more development and consumption that will cause larger GHG emissions unless the world fundamentally alters its dominant development patterns.

Unacceptable Cost of "Carbon Space" and Other Forms of Major Redistributions: The population in Southern nations is three to four times larger than in the North and this disparity is growing. The only way Northern nations could provide enough "carbon space" or "development space" under present atmospheric conditions is to drastically shrink their own economies. While the Northern bloc might be willing to offer a substantial amount of international development assistance, they want to "lift up" the living standards in Southern states without "cutting down" their own economic growth and prosperity, which means they will reject the "carbon space" arguments.

One of the speakers at the conference I attended in India, who was a delegate at the Copenhagen meeting, stressed the "equity" need for reserving enough "carbon space" for India and other developing states with billions of poor people. He then displayed a striking PowerPoint graphic (Figure 5) showing the change in the US economy required to provide the needed "carbon space":

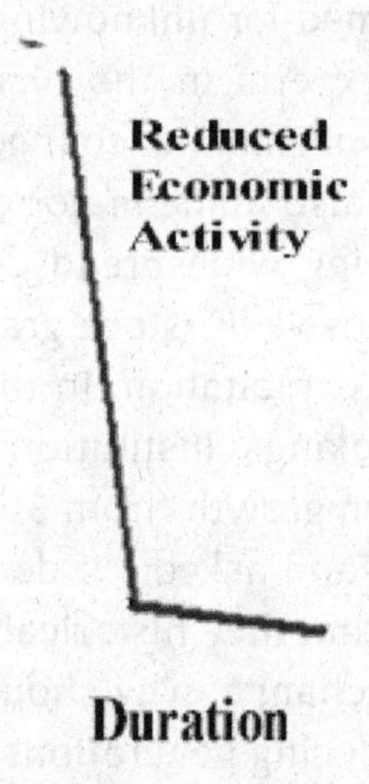

Figure 5: Northern Economic and GHG Cuts Required to Retain Ample "Carbon Space" for Southern States

The speaker appeared quite proud that his controversial plan required such a radical transformation of South–North economic relations, but I do not believe there is the slightest chance that the US and the other Northern bloc countries would really jeopardize their citizens' economic and social welfare to reserve substantial "carbon space" for the developing countries. This is one of the primary "political suicide threshold" issues that no Northern leader could survive if he or she attempts to impose such a devastating economic burden on the nation's own people. Aside from promising appreciable financial assistance for mitigation and adaptation programs in developing states, the Northern representatives in international climate negotiations have never accepted the idea that climate policy should become the lever initiating a wholesale revision of the economic status and standards of living of all the world's nations. Indeed, it would be essentially impossible in the

US and most Northern states for any government leader to advocate this kind of sweeping global redistributional measure.

Some academicians and other intellectuals in a few Northern states have advocated a similar radical strategy entitled "contraction and convergence" (C&C).[297] Under this conception, which has been tentatively enacted into British law but never implemented, the wealthy Northern countries would agree to contract their economies and reduce their GHG pollution discharges progressively until they reach a point at which the per capita GHG emissions of all nations have converged. Because of the increasing population disparities between the South and North, the affluent Northern nations would have to reduce their economic activities and GHG emissions drastically to come anywhere close to enabling a worldwide convergence of economic productivity, resources exploitation, and GHG discharges consistent with the C&C vision. This is another unrealistically *static* position in a *dynamically* changing world because even if the Northern states were able and willing to achieve this radical convergence in the next 20 years or so, the rapidly expanding population and rising aspirations in Southern countries would require a constantly increasing proportion of economic opportunities and per capita GHG discharges rights that the convergence process is supposed to provide, but never will in reality.

The "C&C" conception is only a little less extreme than the "carbon space" idea because it would require international parity in per capita GHG emissions and substantial redistributions of wealth. It is not easy to imagine any American politician telling voters that they must reduce their economic activities and wealth by approximately 90% or more to achieve a convergence with newly uplifted economic and social conditions in dozens of the world's most impoverished countries. Any promise of "equality of economic results" is also incompatible with the basic capitalist economic philosophy adopted to one degree or another by all Northern societies, which doubtless has not escaped the notice of proponents of these sweeping redistributive conceptions. The critical issue here is whether these schemes are politically and economically viable in the North, which does not seem at all plausible to me.

Competitive Disadvantages: In contrast to flights of fancy about developed nations accepting unrealistic requirements for "development space" and "contraction and convergence" initiatives, the US Senate unanimously rejected the Kyoto Protocol a decade ago largely because China and other large GHG-polluting developing nations were not required to cut their GHG emissions commensurately or to select specific target levels for limiting future GHG discharges.[298] This rejection was based in part on concerns about competitive disadvantages that might arise if American businesses have to bear the costs of reducing GHG emissions while competing foreign firms from developing nations do not.

One of the more serious Northern bloc constraints on international GHG pollution control initiatives has been the fear that domestic firms will not be able to compete against products and services from countries that are not required to bear the expenses of cutting GHG emissions.[299] Some experts have suggested that the competitive disadvantage problem could be ameliorated from the Northern perspective by imposing a GHG tax on imports, which has sometimes been called a "border tax adjustment" (BTA).[300]

A border tax adjustment would increase the price of products and services from foreign producers in countries that have not undertaken comparable GHG emissions-reduction commitments. This BTA treatment would undoubtedly pose serious implementation problems[301] and would be deeply resented and opposed by developing countries, which will claim with considerable international law support that they should be entitled to produce goods and services for transnational trade without paying for GHG emissions-reduction programs imposed by other nations.[302] It is likely that BTA conflicts could keep the World Trade Organization dispute panels busy for the next century.

Domestic Redistribution: The UN and the Southern bloc of nations have been using climate change and other disputed international issues as leverage to promote a general agenda constructed around international redistribution programs aimed at poverty alleviation, creating greater economic opportunities for developing states, and improving standards of living and greater equality for all disadvantaged peoples. However, some Northern policymakers question why the Northern bloc should undertake substantial global redistribution efforts before the major developing states make serious efforts to achieve internal redistribution of wealth and economic opportunities among their own populations. China, India, Indonesia, Brazil, and other influential developing states with many millions of poor people also possess wealthy elites and growing middle-class groups that could do a much better job of raising the standards of living of their destitute fellow-citizens if equitable redistribution was really a high priority for them. But most national elites supported by UN officials want developed nations to promote ambitious international redistribution before they make serious attempts to achieve national redistribution. Most developed states have been unwilling to agree to this priority.

Cognitive psychologists should devote close attention to the participants in international environmental negotiations because so many of these people are prone to hear what they want to hear and to believe what they want to believe regardless of the unfavorable reality. After the 15th Conference of the Parties (COP) in Copenhagen ended in disappointment because of the failure to create a binding international climate agreement, the 16th COP held in Cancun, Mexico in winter 2010 was organized with lower expectations.[303] The heads-of-state of most nations did not attend the meeting; an ambitious treaty to control global climate change was not introduced, much less adopted; and every agreement reached was based on voluntary national participation rather than binding international law commitments. In effect, the participating nations promised to do what they were willing to do voluntarily to combat climate change, no more and no less. This negotiating approach enabled all states to agree to respond to climate change and the need for more economic development by doing what they thought was in their own national interest, at the same time that they were not pressured to do what other nations wanted them to do.

According to the UN Environment Programme, these promised actions were among the most notable "Cancun Agreements":

- Industrialised country targets are officially recognised under the multilateral process and these countries are to develop low-carbon development plans and strategies and assess how best to meet them, including through market mechanisms, and to report their inventories annually.
- Developing country actions to reduce emissions are officially recognised under the multilateral process. A registry is to be set up to record and match developing country mitigation actions to finance and technology support from industrialised countries. Developing countries are to publish progress reports every two years.
- Parties launched a set of initiatives and institutions to protect the vulnerable from climate change and to deploy the money and technology that developing countries need to plan and build their own sustainable futures.
- A total of US$30 billion in fast start finance from industrialised countries to support climate action in the developing world up to 2012 and the intention to raise US$100 billion in long-term funds by 2020 is included in the decisions.
- In the field of climate finance, a process to design a Green Climate Fund under the Conference of the Parties, with a board with equal representation from developed and developing countries, is established.
- Governments agree to boost action to curb emissions from deforestation and forest degradation in developing countries with technological and financial support.
- Parties have established a technology mechanism with a Technology Executive Committee and Climate Technology Centre and Network to increase technology cooperation to support action on adaptation and mitigation.[304]

The Cancun Agreements did not resolve the primary conflicts between North and South states, but reiterated them. The developing countries made clear that any mitigation activities they choose to undertake would have to be supported by ample financing from the North. The agreements did not set a higher priority on climate change mitigation than on new development in relatively poor countries. The lump-sum of funding that was ostensibly committed to climate change programs did not indicate which developed states would pay how much, when, and with what "strings" attached. The management of the "Green Climate Fund," for example, was designed to avoid breaking the impasse over the incompatible priorities of the North and South blocs by adopting equal representation. Of crucial importance, the first two agreements clearly indicated that all of the participating nations are free to choose their own climate mitigation measures and timing while they work with other countries only to the extent they voluntarily choose to do that.[305]

Nevertheless, the UNFCCC Executive Secretary Christiana Figueres contended that: "Cancun has done its job. The beacon of hope has been reignited and faith in the multilateral climate change process to deliver results has been restored."[306] She claimed: "Governments have given a clear signal that they are headed towards a low-emissions future together, they have agreed to be accountable to each other for the actions they take to get there, and they have set it out in a way which encourages countries to be more ambitious over time."[307] And Ms. Figueres declared: "This is not the end, but it is a new beginning. It is not what is ultimately required but it is the essential foundation on which to build greater, collective ambition."[308] It would be hard to be more optimistic than this about an international negotiations process divorced from a widely-acceptable international negotiations purpose.

Secretary Figueres and a great many other negotiators doubtless put greater emphasis on the process than on the specific outcomes; and the voluntary and ambiguous character of the Cancun Agreements reflects their desire to fashion some kind of agreement even if it will not play a major role in resolving global climate change problems. Continuing international negotiations will not produce effective mitigation and adaptation measures as long as the major blocs continue to espouse different national interests and priorities. If leading GHG-polluting nations, whether developed or developing states, cannot agree on a fundamental level about what they want, how they will achieve mutually-shared goals, and who should pay for implementing the necessary climate-policy programs, continued international negotiations would be unlikely to achieve genuine climate change success in the next thousand years. We cannot create effective climate policies without first creating compatible national and international interests that will induce major GHG-polluting countries to engage willingly in the necessary collective actions.

The Durban Platform for Enhanced Action

In December 2007, the Conference of the Parties (COP 13) met in Bali to strengthen the UN Framework Convention on Climate Change and the Kyoto Protocol. These negotiations produced the "Bali Roadmap," which supposedly "consists of a number of forward-looking decisions that represent the various tracks that are essential to reaching a secure climate future."[309] This set of provisions was intended to create an ambitious multilateral agreement on stronger climate change mitigation and adaptation actions in the next few years.

The asserted negotiations "breakthrough" at Bali reflected a broad consensus that developed nations should sharply reduce their GHG emissions, and they should also provide substantial technological and financial assistance to help developing countries reduce their GHG discharges while adapting to climate change hazards. Yet, neither goal has been met. The Bali outcomes largely followed the underlying two-track principles in the Kyoto Protocol, but the Bali Roadmap did not identify when, where, how, and how much the developed states would be obligated to cut their GHG pollution, or how much climate-related foreign aid they would have to give to specific developing countries subject to unspecified terms and conditions.[310] In short, the Roadmap imposed different responsibilities on developed and developing states, while ignoring the crucial question of whether conventional emissions-reduction strategies could succeed well-enough, soon-enough to reduce the global climate change risks that all nations must confront today and tomorrow.

At COP 15 in 2009, a few influential nations negotiated the "Copenhagen Accord" behind closed doors on the last day of the meeting. This Accord called for voluntary GHG emissions-reduction pledges from all participating nations and did not mandate any international GHG discharges limits or specific national commitments. For the first time, developed and developing nations were expected to make significant self-selected pledges to reduce their GHG discharges, but they were not required to identify or implement ambitious GHG pollution restrictions.

As a result of the perceived lack of transparency and consensus-based democratic bargaining, the majority of participating countries at the Copenhagen meeting did not accept the Copenhagen Accord but instead merely "noted" that some nations have agreed to a voluntary national emissions-reduction approach.[311] One commentary on the Copenhagen Conference observed: "The Accord included a goal of limiting the rise in global mean temperature to 2 degrees Celsius — thereby rejecting a stronger target of 1.5 degrees proposed by some developing countries — but contained no concrete commitments by any parties that would suggest that this aspirational goal will be achieved."[312]

In 2010, the "Cancun Agreements" adopted at COP 16 in Mexico were praised in contrast to the Copenhagen Accord because, as the UNFCCC Executive Secretary, Christiana Figueres, contended: "Governments have given a clear signal that they are headed towards a low-emissions future together, they have agreed to be accountable to each other for the actions they take to get there, and they have set it out in a way which encourages countries to be more ambitious over time."[313]

The rather nebulous Cancun Agreements mainly adhered to the two-tier pattern of emissions-reduction patterns defined by the Kyoto Protocol, and the Agreements also promised a substantial amount of financial assistance from affluent states to developing countries.[314] The emissions-reduction process recommended at the Cancun meeting did not incorporate specific numerical pollution control targets applicable to particular nations, and instead retained the voluntary emissions reductions pledged in response to the Copenhagen Accord.[315]

The Cancun Agreements created a number of international mechanisms aimed at promoting assistance for developing nations, including a "Registry" to match the developing countries' mitigation actions with suitable "finance and technology support" from affluent industrialized states; a "Green Climate Fund" to provide $30 billion in "fast start finance" from wealthy nations that would initially "support climate action in the developing world up to 2012," with "the intention to raise $100 billion [annually] in long-term funds by 2020"; a general promise to "curb emissions from deforestation and forest degradation in developing countries receiving technological and financial support"; and a "Technology Executive Committee and Climate Technology Centre and Network to increase technology cooperation to support action on adaptation and mitigation."[316]

Unfortunately, none of these idealistic "Cancun Agreements" has yet to materialize in concrete form. The UNFCCC Executive Secretary asserted: "This is not the end, but it is a new beginning. It is not what is ultimately required but it is the essential foundation on which to build greater, collective ambition."[317] In short, 18 years after the adoption of the UNFCCC treaty, the multilateral negotiating process was still aiming for "a new beginning."

The next annual meeting, the Durban Conference in 2011, drew more than 12,000 people to participate in, or report on, the joint negotiations for the 17th COP of the UNFCCC and the 7th Meeting of the Parties to the Kyoto Protocol (CMP 7).[318] In addition to these primary international agreements, there were also influential dialogues in several Working Groups and Subsidiary Bodies set up by the UN to address specific climate change problems. This complex and often controversial process yielded 19 formal COP decisions, 17 CMP decisions, and a number of significant decisions by the lesser advisory bodies.[319] The most important consequences of the Durban meeting were an agreement for a second commitment period of the Kyoto Protocol and the promulgation of the Durban Platform for Enhanced Action, which is intended to create a new binding legal commitment applicable to all participating countries and to strengthen the pledges and processes adopted at the previous COP meetings.

The International Institute for Sustainable Development's (IISD) daily reports from Durban[320] and other topical commentaries[321] provide detailed coverage of the many Durban Conference sessions, debates, decisions, and unresolved disputes. I therefore intend to discuss here only two core themes: the continuing mitigation focus on conventional back-loaded emissions-reduction commitments despite the previous failures, and the continuing conflicts among divergent national interests and priorities that have not been effectively overcome by a barrage of utopian phraseology.

The first paragraph in the Durban Platform emphasized the recognition of the parties:

> that climate change represents an urgent and potentially irreversible threat to human societies and the planet and thus requires to be urgently addressed by all Parties, and acknowledging that the global nature of climate change calls for the widest possible cooperation by all countries and their participation in an effective and appropriate international response, with a view to accelerating the reduction of global greenhouse gas emissions....[322]

Despite this call for "the widest possible cooperation," the Durban meeting reflected sharp differences in the negotiating positions of various blocs of nations. China, India, and many other developing nations insisted on the creation of a second commitment period for the Kyoto Protocol in order to retain the two-track process for emissions-reduction and funding obligations imposed on the Annex I developed states in comparison to the relative lack of commitments of developing nations. China and other non-Annex I nations emphasized the need for the preservation of the "common but differentiated responsibilities" doctrine in the Kyoto treaty in order to ensure that no specific pollution control obligations would be required or expected from still-developing states.

In contrast, a number of Annex I nations aside from the European Union rejected continuation of the two-track Kyoto Protocol practices because some developing countries with rapidly growing economies had become among the world's largest GHG dischargers despite their constant refusals to accept any global emissions-reduction commitments. Canada, Japan, and Russia announced that they would not agree to a second commitment period as long as this bifurcation among the major GHG dischargers was maintained,[323] and the United States continued to reject the Kyoto treaty, which it had never ratified.[324] The EU adopted the political position that they would only support a second commitment period under the Kyoto treaty if all UNFCCC member nations, including developed and developing states, agreed to produce a new legal mandate cutting worldwide emissions substantially by 2020 without retaining the two-track process.[325]

After considerable in-fighting, the various blocs agreed to a compromise agreement meant to satisfy all divergent positions to some degree: This asserted "historical breakthrough"[326] would combine a second commitment period for the Kyoto Protocol extending until either 2017 or 2020[327] — the choice of the second commitment period's final date was deferred until the 2012 CMP 8 meeting in Qatar. The parties would also create a new "protocol, legal instrument or agreed outcome with legal force under the Convention applicable to all Parties" that should be adopted no later than 2015 and implemented by 2020.[328] This future single-track emissions-reduction and funding agreement "with legal force" is the core focus of the Durban Platform for Enhanced Action. Numerous developing states contended that their ability to participate effectively in meeting the requirements adopted under the Durban negotiations will be determined by the amount of financial and technological assistance they receive from affluent developed nations and multilateral funding sources.[329]

It is difficult to understand how anyone could believe that the two distinctive emissions-reduction processes created at the Durban meeting, the Kyoto Protocol extension and the single-track agreement with legal force, could achieve any genuine climate mitigation progress. With regard to the Kyoto second commitment period, the world's three largest GHG-polluting nations, China, the US, and India, will not be covered by the revised emissions-reduction standards because two of them are non-Annex I states and the US never joined the Kyoto treaty. Some other developed nations with substantial GHG discharges have chosen to withdraw from the Kyoto treaty, or have threatened to do so, as a result of the perceived unfairness and competitive disadvantages of the two-track approach. It is also important to stress that more than 150 developing nations and countries with emerging economies are not included in Annex I and therefore have no specific obligations for meaningful GHG reductions. When all of the exceptions and loopholes are aggregated, the prospect of the revised Kyoto second-commitment emissions-reduction standards helping to stabilize or curtail the GHG concentration in the atmosphere is virtually non-existent.

Although the specific GHG discharges reductions that Annex I states will be required to make during the second commitment period have not yet been agreed upon, there was frequent discussion at the Durban meeting of adopting a 25–40% cutback by 2020 "as considered in the IPCC's fourth assessment report."[330] However, the IPCC's estimate that there must be *at least* a 25–40% reduction in *worldwide* GHG discharges to keep the expected temperature increase below 2 degrees Celsius is much more stringent than the imposition of these percentage emissions-reduction rates only on Annex I nations that now put out less than half of the annual global GHG discharges.

From the perspective of this book, the 25–40% emissions-reduction rates that may be adopted for the second Kyoto commitment period are a striking example of how the leading climate negotiators and their expert advisors do not understand the "stocks and flows" character of climate change processes or the "reducing the increases" problem undermining effective mitigation actions. Let us suppose, as a hypothetical, that the range of Annex I cutbacks between 25% and 40% of 1990 GHG emissions is adopted next year and is properly implemented with transparent MRV. Under this scheme, the Annex I states will continue to discharge the remaining 60–75% of their 1990 discharges, and the non-Annex I countries (including the non-member US) will be allowed to continue "business as usual" GHG emissions that will almost certainly be even higher than the Annex I nations' emissions.

These huge residual GHG discharges under the revised Kyoto Protocol unarguably will increase the concentration of GHGs in the atmosphere every year while the treaty remains in effect, and they will cause climate change dangers to grow worse each year. There will be absolutely no climate benefits from the wholly inadequate emissions reductions in the second Kyoto commitment period except for the lame argument that climate change hazards would be slightly worse if the participating Annex I nations made no GHG pollution cuts at all. It is vital to recognize that the GHG concentration in the atmosphere is already too high and is causing a wide range of increasing harms; if we allow a large volume of additional GHGs to expand the atmospheric "stock" of greenhouse gases every year, as the two Durban emissions-reduction programs would do, climate policymakers realize that these feeble mitigation efforts are steadily making climate conditions worse, no matter how sincere the negotiating efforts may be.

The widely-accepted choice at the Durban meeting to "develop a protocol, legal instrument or agreed outcome with legal force under the Convention applicable to all Parties" does not ensure that this one-track legal agreement will be any more stringent than the emissions-reduction standards set in the second commitment phase of the Kyoto Protocol. There is no indication that the negotiating process for this new multilateral agreement will be able to harmonize the conflicting national interests of various blocs of participating nations, rather than producing a lowest-common-denominator set of mitigation and adaptation obligations that will achieve virtually nothing in responding to increasing climate change dangers.

Although the Durban Platform for Enhanced Action is supposed to be completed by 2015 and implemented by 2020, there was no explanation of why these stages were deferred for almost a decade and there was no definition of what "enhanced action" will mean in practice. The Platform text did provide "that the process shall raise the level of ambition and shall be informed, inter alia, by the Fifth Assessment Report of the Intergovernmental Panel on Climate Change, the outcomes of the 2013–2015 review and the work of the subsidiary bodies."[331] However, the availability of more recent IPCC findings and Working Group negotiations offers no assurance that the new legal agreement will impose a draconian mandate for ambitious mitigation because there is already ample information documenting climate change risks and current harms without inspiring suitable mitigation actions.

The Platform also provided that the negotiators should launch a workplan "that can close the ambition gap with a view to ensuring the highest possible mitigation efforts by all Parties."[332] But there is no reason to suppose that some of the large GHG-polluting developing countries will interpret this phrase as requiring them to subordinate further development plans to mitigation efforts. At the Durban meeting, "Brazil, China, India and other developing countries argued that agreeing [to] a second Kyoto period was a legal and moral obligation of industrialised countries and not a bargaining chip to extract concessions from developing countries."[333] The fact that they did eventually recognize there would be no Kyoto second commitment period without their acceptance of a subsequent one-track mitigation mandate does not imply that they will engage in the "highest possible" efforts.

According to the IISD Summary of the Durban proceedings, the parties emphasized the language in the UNFCCC recognizing "that social and economic development and poverty eradication are the first and overriding priorities of developing countries."[334] The host nation, South Africa, stated that the instrument must reflect "the need to capture and support different kinds of effort in a common framework" and must support "the variable efforts of countries at different points on the development spectrum, thus respecting while recasting the Convention's principle of common but differentiated responsibilities."[335] In order to attain sufficient support for the proposed comprehensive legal instrument or outcome with legal force, the lead EU representative assured "China and India that they would simply be expected to turn their Cancun pledges into new legal arrangements,"[336] which was a disastrous concession wholly inconsistent with promoting the "highest possible" mitigation efforts.

The bottom line is that some commentators have praised the Durban Platform as the beginning of new comprehensive negotiations based on a presumption that all GHG-polluting nations will contribute to climate change mitigation; while critics of the Durban negotiations have pointed to the absence of specific emissions-reduction targets, the additional delays imposed by decisions deferred until later COPs, and the lack of more ambitious mitigation commitments.

For example, one assessment noted: "Those who see Durban as a bitter disappointment point out that smokestacks and automobile exhaust pipes continue to spew carbon, that atmospheric warming will continue apace, and that the world leaders still haven't actually done anything."[337] Yet, the same commentary concluded: "Durban's real accomplishment was to keep the slow, torturous process of climate negotiations alive, with the biggest carbon emitters now involved."[338] This dichotomy was also emphasized by a German assessment that "progressive countries score a realpolitik victory in Durban while the real climate continues to heat up."[339]

In my opinion, the Durban Conference and the past UNFCCC meetings could not be successful because they have been unable to overcome the markedly different goals and priorities of developing and developed nations, as well as the goals of other more limited international blocs. I must therefore agree with the negative conclusion of William R. Moomaw, a professor of international environmental policy:

> The climate negotiations just ended in Durban completed two decades of frustrating, failed diplomacy. The US and China continue to hide behind one another, each trying for an elusive short-term economic advantage and creating a catastrophic risk for us all. It is time to admit that we are negotiating the wrong treaty. The [Durban] "agreement" that sometime in the future all nations would need to reduce emissions was begrudgingly agreed to and clearly fails to meet the interests of a majority of nations.[340]

If we cannot find a way to bridge the gaps in perceived national interests, it is hard to see how new beginnings in climate negotiations that have proved fruitless for the past decade will yield meaningful mitigation achievements in the next decade or ever.

Identifying a Potential Solution

The Southern nations want to increase economic and social development while forcing the North to make the great majority of

GHG emissions reductions and to pay for whatever mitigation and adaptation programs the Southern bloc states do choose to undertake. The Northern nations want to retain their relatively affluent standards of living while requiring all major GHG-polluting Southern nations to engage in mitigation efforts that reduce climate change risks. The North also wants to ensure that funding they provide for international climate-control programs is used effectively for that purpose and not to create anti-competitive subsidies.[341] These conflicting climate-policy objectives support my conclusion that the core international disagreements cannot be resolved through further negotiations based on the familiar processes, positions, and national interests that have repeatedly failed to attain meaningful commitments toward combating global climate change risks.

A successful solution will require cooperative actions by all of the major GHG-polluting nations, including developed and developing states, but the world's nations are not even close to reaching meaningful agreements because of underlying long-term conflicts in national interests and priorities.[342] No international laws would allow the North to force non-participating sovereign states of the South to reduce their GHG emissions by specified amounts; and no international laws would prevent the developing nations from focusing much more on economic and social welfare growth than on climate control. In the meantime, global climate conditions are steadily growing worse and may be approaching disastrous "tipping point" changes that cannot be reversed.[343]

Climate change mitigation advocates must recognize that they cannot succeed by relying on the same ineffectual measures and incompatible arguments that have been consistently failing for decades. An ambiguous international legal agreement, such as the Cancun Agreements, that avoids rather than resolves climate-policy and sustainable development conflicts could not be worthwhile. Indeed, a treaty that masks or avoids the critical disagreements may be actively harmful by creating the illusion of cooperation and progress without any commensurate accomplishments in practice. If nebulous idealized principles without any specific mitigation commitments could overcome atmospheric greenhouse effect and global warming problems, the world undoubtedly would already be safe from future climate change harms.

There is no international consensus on the trade-offs between expanding development and GHG emissions reduction. For example, China and India, with about a third of the world's population, still have hundreds of millions of poor people despite remarkable economic progress in the past two decades. These nations rely extensively on coal-burning power plants to generate ever-increasing amounts of electricity and ever-increasing amounts of GHG pollutants because they lack oil and natural gas deposits; and coal, which they do have, is the least expensive fossil fuel that can meet their growing energy demands. Both nations plan to continue expanding their reliance on coal-combustion technologies, which is bound to be a GHG pollution disaster further compounding global climate change problems.[344]

The Chinese and Indian governments unquestionably understand that their reliance on "dirty" coal combustion for energy production and other GHG-producing industrial practices will substantially worsen global warming and will cause significant harms to their own regions and peoples. Yet, the governments of both nations regard increasing economic and social development as their highest priority and they have refused to accept any numerical limits on their GHG discharges or to promise any specific future emissions reductions.[345] In essence, they will not compromise their drive for greater development by agreeing to restrain GHG discharges from increasingly dangerous production and consumption activities, except for limited "carbon intensity" improvements.

Most developing nations will not accept pollution restrictions that preclude exploitation of the "dirty" energy-producing and manufacturing technologies currently necessary to expand their economic growth. These nations regard any request to sacrifice increased development as another attempt by wealthy states to prevent less advantaged countries from "catching up" to a comparable level of economic and social welfare they feel entitled to attain. As long as the developing nations follow in our "dirty" economic development footsteps by using the same GHG-polluting industrial and consumption methods that the affluent nations created,[346] any efforts to mitigate climate risks will be hopeless and more interminable international meetings on climate change will be pointless.

At the other extreme, the US Senate rejected the Kyoto Protocol a decade ago and since then has refused to adopt any climate change legislation on the grounds that it might increase the costs of energy and transportation for American consumers, and might threaten American jobs created by fossil fuel production and distribution. The Bush Administration also claimed that voluntary GHG pollution control initiatives would prove sufficient,[347] a position most climate experts regard as nonsense.

To summarize, most developing states will not accept specific GHG emissions-reduction commitments that may impede greater economic growth; while the Northern countries will not accept redistributive measures that threaten their economic and social welfare. The North also will not provide funding for climate programs without imposing "strings" they deem appropriate but the Southern states consider intrusions on their sovereignty.

My solution to this impasse is straightforward: We must develop climate programs that can accomplish what *both blocs want most*. To the extent that developed nations genuinely want to overcome climate change problems, these nations must change their focus or frame of reference away from imposing emissions-reduction targets by asking how they can persuade developing countries to reduce their GHG discharges. Climate change cannot possibly be contained or overcome unless both blocs of countries collectively undertake effective pollution control measures, which means mitigation proponents must gain the cooperation of nations with divergent interests and priorities.

To reach a mutually satisfactory agreement, the developed nations and multilateral institutions must modify their climate mitigation strategy to include fostering greater development in less affluent nations while at the same time reducing GHG pollution from as many sources in as many places as practicable. Proponents of effective precautionary programs must be able to accomplish both *mitigation and development*. The only realistic course of action that meets both requirements is offering the developing nations *clean* GHG-free technologies that will enable them to increase their prosperity without discharging large quantities of GHGs. We must also offer them substantial development funding and technology transfers tied to the deployment of GHG-free technologies and processes. In other words, they should be offered foreign assistance to support "clean" development efforts but not "dirty" development practices.

Even meeting stringent GHG emissions-reduction targets in Northern nations — perhaps an 80% decrease in GHG pollution within the next three decades — would not offer developing countries the economic growth and improved social welfare they want. If wealthy states gave the developing nations improved emissions-reduction technologies, such as lower-polluting vehicle engines, these technology transfers would help cut GHG emissions to some extent compared to unrestricted BAU pollution, but they would not significantly promote the greater development that most poor countries insist is their highest priority. In contrast, the provision of GHG-free replacement technologies, especially affordable renewable energy technologies, could promote development and eliminate GHG pollution at the same time. A new carbon-free electric power plant, for example, can provide the energy required for further economic development without discharging residual GHGs that would remain in the atmosphere for centuries after the adoption of inadequate GHG emissions-reduction programs.

The frustrating fly-in-the-ointment is that the developed nations have not yet designed and disseminated sufficient clean replacement technologies that could achieve improvements in the economic welfare of developing nations without producing harmful GHG emissions. We do not have enough of the clean technology that they need, and we need, for there to be a realistic chance of effectively reducing the GHG concentration in the atmosphere. There is no benefit in conducting expensive, time-consuming, ultimately pointless international climate conferences when we lack the fundamental requirements for worthwhile climate-control agreements.

One of the Cancun Agreements called for the creation of a Technology Mechanism with two components: a Technology Executive Committee with 20 members appointed from different countries, which is intended to "promote and facilitate collaboration on the development and transfer of technologies for mitigation and adaptation between governments, the private sector, non-profit organizations and academic and research communities"[348]; and the Climate Technology Centre and Network, which was established to provide "advice and support related to the identification of technology needs and the implementation of environmentally sound technologies, practices and processes."[349] At the Durban meeting the following year, the main topics of discussion on the Technology Mechanism were the decision on which nation would serve as the host for the Climate Technology Centre and what sources of financial support could be accessed to begin funding various technology-related programs.[350] While there were many references to technology transfers and to the assessment of adaptation technology needs for developing nations, there appeared to be little or no focus on the development of carbon-free replacement technologies that could benefit both developed and developing countries.

Although the Technology Mechanism could eventually prove useful, it is still largely inchoate and ineffectual. One primary goal drawn from the UNFCCC text is to "promote and cooperate in the development, application and diffusion, including transfer, of technologies, practices and processes that control, reduce or prevent anthropogenic emissions of greenhouse gases not controlled by the Montreal Protocol in all relevant sectors, including the energy, transport, industry, agriculture, forestry and waste management sectors."[351] This passage is filled with so many general objectives that it is impossible to determine where the main Technology Mechanism efforts will be applied. However, the major aim of this statement seems to be to facilitate all dimensions of pollution control and adaptation efforts including emissions-reduction programs. The phrase "prevent anthropogenic emissions" has not been explained or implemented in a detailed manner, and it certainly has not produced any substantial contribution to reducing climate change hazards.

The Kyoto Protocol second commitment period and the Durban Platform one-track legal agreement were much more central to the Durban proceedings than the Technology Mechanism. It is indisputable that conventional emissions-reduction programs are still the primary mitigation approaches in UNFCCC proceedings while clean GHG-free replacement technologies have been attracting relatively little attention. In contradiction to the consensus mitigation approach, this book means to show that GHG emissions-reduction programs, as consistently advocated by UN initiatives, are too feeble and too slow to attain significant climate change progress. Apparently, Professor Moomaw has a similar perspective because he contends that we must "begin by replacing the pollution model of controlling carbon emissions through targets and timetables with a new mutual gains economic development model that establishes a right of universal access to clean, low-carbon energy services."[352]

After developed nations design and refine innovative GHG-free replacement technologies that could be transferred to developing countries without requiring them to make economic sacrifices, which they will refuse to do, numerous international meetings would be necessary to determine how much financial aid and technology transfer funding assistance should be provided by the affluent nations. At that time, the technical ability to eliminate or drastically reduce GHGs while enabling increased development will be demonstrated, as it has not been today, and the debate would turn to which nations should pay how much to combat global climate change.

If effective GHG-free replacement technologies can be devised and widely deployed in the next few decades, developing states would gain the economic growth and social welfare enhancements they want while the developed nations would obtain climate change progress and clean economies without having to give up anything but a modest amount of capital and some technological export profit opportunities. In contrast, the current demands from the US and other Northern states for quantitative emissions-reduction commitments by the developing nations will continue to be ignored because they conflict with comparatively poor states' self-improvement goals and rarely promote greater economic and social welfare.

We need to design and disseminate clean energy and transportation technologies as promptly as feasible. Transfers of interim emissions-reduction technologies, such as dual-fuel vehicle engines, may be better than nothing, but they cannot reduce GHGs nearly as effectively as all-electric or hydrogen fuel-cell cars, and they will not be attractive to most developing nations because they do little to promote greater prosperity. Until the developed nations have produced affordable GHG-free replacement technologies, the transition to a GHG-free economy that can simultaneously facilitate increased economic development in less affluent states will be impossible and so will any meaningful international negotiations agreement.

I must disagree with Dr. Hansen and his preference for "an across-the-board fee on fossil fuel carbon [able to] work on a global basis, in a way that is fair, because unless there is a universal carbon fee, it will be ineffective."[353] In its brief discussion of developing countries and their asserted willingness to limit GHG emissions, Hansen's book does not explain how a fee-and-dividend program would facilitate economic growth in developing states, though that is their highest priority. Instead, Hansen's explanation seems to suggest that climate change control is what all nations really want most, and his fee-and-dividend proposal is the best way to get there. Here is his wishful-thinking explanation of why China would adopt the fee-and-dividend system:

> The key requirement is that the United States and China agree to apply across the board fees to carbon-based fuels. Why would China do that? Lots of reasons. China is developing rapidly and it does not want to be saddled with the fossil fuel addiction that plagues the United States. Besides, China would be hit at least as hard as the United States by climate change. The most economically efficient way for China to limit its fossil fuel dependence, to encourage energy efficiency and carbon-free energies, is via a uniform carbon fee. The same is true for the United States....
>
> Agreement between the United States and China comes down to negotiating the ratio of their respective carbon tax rates. In this negotiation the question of fairness will come up — the United States being more responsible for the excess carbon dioxide in the

> air today despite its smaller population. That negotiation will not be easy, but once both countries realize they are in the same boat and will sink or survive together, an agreement should be possible.[354]

Dr. Hansen's explanation does not even mention that China's foremost national goal is to expand economic development for the benefit of its elites as well as hundreds of millions of still-poor people. It would be interesting to behold Hansen's "same boat" negotiations in light of the competitive disadvantages it could confer on one nation or another, but these negotiations are extremely unlikely to occur as long as the conflicting national interests of China and the United States remain in place. It appears that Dr. Hansen believes the two countries share the same interests and priorities, which is not a realistic assessment with regard to climate change policies or much of anything else.[355]

China's behavior in recent years emphatically shows that it does not consider itself in the "same boat" as the US.[356] If China really subscribed to Dr. Hansen's idealistic arguments, it would have tried to work out an international GHG pollution control deal instead of sabotaging more ambitious mitigation proposals at the Copenhagen Conference.[357] China wants the US and other Northern states to make major emissions-reduction commitments while it continues to be the world's largest GHG polluter and coal user, building new coal-burning power plants every week that will lock in vast amounts of fossil fuel GHG pollution for decades to come.[358]

It is true that China has been investing large sums on research for creating cleaner technologies, which it wants to sell to other countries and also will use to reduce its carbon dioxide emissions whenever convenient and affordable, but China is still firmly on the path to expanding coal utilization for many years.[359] One recent news article reported: "Even as China has set ambitious goals for itself in clean-energy production and reduction of global warming gases, the country's surging demand for power from oil and coal has led to the largest six-month increase in the tonnage of human generated greenhouse gases ever by a single country."[360]

Another recent news article discussing China's impact on climate change conditions quoted a statement from Mark Lynas, a British environmental writer, TV show host, and advisor to the government of the Maldives:

> The NGO [non-governmental organization] movement is ten years out of date. They're still arguing for 'climate justice', whatever that means, which is interpreted by the big developing countries like India and China as a right to pollute up to Western levels. To me carbon equity is the logic of mutually assured destruction. I think NGOs are far too soft on the Chinese, given that it's the world's biggest polluter, and is the single most important factor in deciding when global emissions will peak, which in turn is the single most important factor in the eventual temperature outcome.
>
> Too many leftist activists are therefore tending to side with the big polluters because they think they're standing in solidarity with the world's poor. I far prefer President Nasheed's (of the Maldives) take that "two wrongs don't make a right." In his plenary speech, he said: "We don't want carbon, we want development. We don't want coal, we want electricity. We don't want oil, we want transport." This seems to me to be the only logical path forward at this stage in the game....
>
> I think the bottom line for China (and India) is growth, and given that this growth is mainly based on coal, there is going to have to be much more pressure on China if global emissions are to peak within any reasonable time frame. In Beijing the interests of the Party come first, second and third, and global warming is somewhere further down the list. Growth delivers stability and prosperity, and keeps the party in power....
>
> [W]estern electorates will not buy a deal that does not involve some meaningful action by China. "Why should I do anything when China is building X new coal-fired power stations each week," is the most common response I've come across, and is actually a fair point.[361]

The comment by the Maldives President that they want more development, electricity, and transportation, not more fossil fuels, is entirely compatible with the GHG-free replacement-technology approach I am proposing: We must replace environmentally harmful technologies with GHG-free processes that can attain the same welfare goals without continuing to degrade climate conditions. This is the only way I can see to meet the separate distinctive needs and priorities of the South and North. Unfortunately, while the Northern developed nations continue to dither and disagree, China and India are locking in decades of increased GHG discharges from a constantly increasing number of new coal-fired power plants.

No one could claim that it would be easy or inexpensive to replace the current Northern GHG emissions-reduction strategy with GHG-free or very-low-GHG replacement technologies, but the consensus emissions-reduction programs can never succeed on an international plane because they do not help provide the greater economic growth and social welfare that the large GHG-polluting developing nations want above all else. The consensus emissions-reduction programs were (badly) designed to cut greenhouse gases, not to expand economic and social development in less affluent countries. The emissions-reduction programs of the North are only creating the illusion of climate change progress, and not much progress at that, while they waste precious time and valuable regulatory resources that could be used to support the transition to a GHG-free economy and society to a much greater extent.

Continued reliance on a climate mitigation strategy that cannot succeed in overcoming or even significantly reducing climate change risks is unarguably a major climate-policy failure that can only be remedied by becoming more responsive to the needs and priorities of all major GHG-polluting countries, including developed and developing states. The best and probably the only realistic way to achieve that is by creating clean GHG-free replacement technologies, which can eliminate the worst sources of GHGs while nonetheless enabling greater prosperity and new opportunities for economic growth that the developing nations urgently want and that their expanding populations are bound to demand.

It is most unfortunate that the great majority of governments remain fixated on inflexible perceptions of their national interests despite many severe natural catastrophes in 2010 and 2011, which cost thousands of lives and billions of dollars in property losses.

Recent floods, droughts, heatwaves, tornados, cyclones and hurricanes, as well as other extreme weather conditions, cannot be attributed to one specific cause, but as a statistical group they clearly meet scientific predictions of increasing climate change disasters. Yet, the large GHG-polluting nations that are among the most vulnerable to climate-related damages, including China, India, and the US, have been doing little or nothing to create mutually-supportive international mitigation initiatives.

CHAPTER V

Overlapping Institutional Responsibilities

Overcoming global climate change will probably be the most difficult, expensive, complicated, controversial task the human race has ever undertaken on a collective basis, and there is certainly no guarantee that we will succeed in meeting this responsibility to the Earth and to future generations of human beings. Successful climate change policies must accomplish two demanding goals concurrently: eliminating as much residual GHG pollution as feasible to stabilize and then reduce the atmospheric GHG concentration that causes the greenhouse effect, and promoting greater economic and social welfare in developing countries that otherwise will continue discharging more GHGs into the atmosphere every year. Regrettably, the world's leaders go on debating whether to reduce GHG pollution 20%, 30%, 50%, 60%, or 80% by various future target dates, while relying on consensus emissions-reduction programs, including cap-and-trade systems and carbon offset programs, that cannot achieve *either essential function*.

Instead of pursuing conventional "solutions" that cannot attain any significant climate change progress, we need to build political support for minimizing GHG discharges through a GHG-free replacement-technology strategy that is much more likely than an emissions-reduction approach to reduce climate change risks under real-world conditions. Concerned countries could impose draconian pollution controls that would threaten the economic prosperity, employment levels, and lifestyle choices in their nations, but this drastic regulatory strategy could not avoid damaging the economies of Northern and Southern states and also could not offer the developing countries what they want most in return for substantially cutting their GHG emissions. From a practical perspective, it is hard to imagine how developed or developing nations could attain major GHG reductions without creating feasible GHG-free replacement technologies. My climate-policy position is that we need to put the great majority of our mitigation efforts and investments into conducting this GHG-free replacement-technology transformation as widely and rapidly as practicable, rather than deferring it for decades while we employ expensive emissions-reduction programs that will yield no meaningful climate benefits.

It is easy to understand why many politicians would prefer to adopt back-loaded emissions-reduction programs that will impose the strictest regulatory requirements and greatest expenses on future generations. The current leaders would not have to worry about being held accountable for the non-results of their mitigation initiatives, or about public resistance to costly emissions-reduction efforts that will require many industrial and consumption behavioral modifications without yielding comparable benefits. The leaders would not have to worry about resolving important scientific and economic uncertainties based on less than complete information. And the leaders also would not have to jeopardize their campaign contributions by leading the fight against powerful fossil fuel industries and other influential financial interests.

Implementing a thoughtful climate change precautionary program in the United States, a country that is involved in two costly wars, that has not yet rebounded from a severe financial crisis, that has a higher unemployment rate than at any time since the Great Depression of the 1930s, that has accumulated a higher national debt than ever before except perhaps during World War II, that has been losing its competitive economic and educational edge and its sense of security compared to other nations, and that has largely accepted a divisive, polarized, dishonest political system in which virulent opposition often arises simply for the sake of opposition, may realistically be too much for the US political agenda to bear in the next few years. It would be hard to find a recent era in which the political and economic conditions were less amenable to promoting environmental protection initiatives. This unpleasant but unarguable reality presents a crucial dilemma: We must undertake effective, urgent actions to prevent irreversibly expanding many climate change risks, and yet the present socio-economic conditions do not appear in the least propitious for this necessary struggle. The familiar adage of being caught "between a rock and a hard place" is definitely applicable.

Nevertheless, climate protection advocates must try to undertake *useful* mitigation efforts because pushing climate change prevention goals into the indeterminate future is sure to cause more severe climate-related disasters. Yet, unlike many climate experts, I do not believe we would benefit from making substantial mitigation commitments for the sake of doing *something now* without being reasonably sure that the measures we choose will actually prove useful.

Adopting Initial or Interim Mitigation Measures

Many prominent environmental advisors and environmental NGOs maintain that we must enact climate change legislation, such as the Waxman–Markey or Kerry–Lieberman Bills, as soon as possible even if the resulting regulations would be very weak and rife with special-interest loopholes. For example, the Nobel Prize-winning economist and columnist, Paul Krugman, claimed: "After all the years of denial, after all the years of inaction, we finally have a chance to do something major about climate change. Waxman–Markey is imperfect, it's disappointing in some respects, but it's action we can take now. And the planet won't wait."[362]

I disagree with this notion that we must do *something now*, no matter how ineffectual and wasteful. Much of this book has been devoted to showing that the consensus GHG emissions-reduction programs and cap-and-trade systems at the heart of the recent congressional bills, Administration proposals, and other pollution-control initiatives will *not* "do something major about climate change," to use Dr. Krugman's phrase. Instead, they will only waste scarce resources and time while achieving no tangible climate change benefits.

The emissions-reduction provisions in this legislation will yield no climate benefits, not merely less progress than we would like, because the Waxman–Markey Bill and similar consensus proposals would create "reducing the increases" outcomes that ensure the atmospheric GHG concentration and greenhouse effect will continue to grow worse. As a respected economist, Professor Krugman must recognize that the proponents of climate change legislative bills should assess the opportunity-costs, or foregone opportunities, of interim mitigation efforts that may be wholly based on wishful thinking and might achieve nothing more than political hype, wasted resources, and irreplaceable lost time.

I do not agree with the view that we must accept ineffective political, economic, or technological measures because *we have to do something now* even when these measures are inadequate to achieve any genuine climate change progress. How many real chances will we have in the next decade to create effective climate mitigation legislation in the US, and how many of those scarce opportunities will be squandered by accepting ineffectual emissions-reduction programs because "the planet won't wait," to use Dr. Krugman's misguided phrase?

The "we must do something now" mantra is bound to require substantial expenditures in response to the complexity and difficulty of climate change problems; and it seems nonsensical to support these investments with no assurance that they will produce any meaningful benefits. Making a harmful climate condition a bit less *bad*, but not even close to *good* — the central weakness of "reducing the increases" mitigation programs — will be counter-productive due to the losses of money, time, regulatory resources, public support, and administrative morale. We cannot afford to waste billions of dollars on failing climate mitigation plans that will foreclose better precautionary measures because undertaking *something now* may never produce actual climate progress and instead may only misuse limited funding and resources.

When the Senate rejected the Lieberman–Warner Climate Security Act in June 2008, a *New York Times* editorial described this bill as "short of what most climate scientists believe is necessary but an important first step."[363] Again, I disagree: *A first step is only valuable if it is heading in the right direction,* and this idealistic editorial failed to consider several important reasons why recent attempts to implement first-step or interim emissions-reduction measures may have proven more harmful than helpful.[364]

The technology needed to attain an interim GHG emissions-reduction target will often be entirely different from the technology needed to meet a carbon-free performance standard. Motor vehicles, for example, can reduce GHG discharges by using less powerful engines and lighter materials, by substituting biofuels for high-CO_2 gasoline, or by using hybrid engine designs.[365] Cars with these interim modifications will nevertheless continue to discharge large amounts of persistent residual GHGs into the air and cannot approach the GHG-free cleanliness of all-electric or hydrogen fuel-cell vehicles. It is important to recognize that the money and efforts invested in this kind of initial or interim fossil fuel-based fuel-efficiency technology will no longer be available to support the transition to GHG-free transportation technologies.

As a case in point, General Motors (GM) ran a detailed series of online advertisements in May 2008 that said their employees also live on planet Earth and they are working concurrently on several new ways to reduce the global warming pollution from their vehicles. The GM efforts included engine and materials improvements, modified diesel engines, flex-fuel engines using a low percentage of ethanol, more ambitious biofuel designs, hybrid gas and electric engines, all-electric cars, and hydrogen fuel-cell vehicles.[366] The GM ad did not predict specific dates for the widespread deployment of any of these alternative vehicle efficiency improvements, but it did indicate that GHG-free designs are bound to be quite expensive and will not be broadly available until a long time in the future, if ever.[367]

In 2010, GM announced that the Chevrolet Volt, a plug-in electric hybrid, would be available the following year with a roughly $41,000 price tag (before tax incentives) for a compact car that can only travel about 40 miles on electricity before it must plug in or use its gasoline engine to recharge the batteries. One reporter's negative assessment noted that the Volt "requires premium gasoline, seats only four people (the battery runs down the center of the car, preventing a rear bench) and has less head and leg room than the $17,000 Chevrolet Cruze, which is more or less the non-electric version of the Volt."[368] American motor vehicle designers undoubtedly could do much better than the Volt if they were given access to adequate research and development resources.[369] Predictably, GM has not yet accepted overcoming climate change dangers as one of its corporate objectives or responsibilities.

GM and other American automobile manufacturers have been putting more money and engineering efforts into interim design modifications that can meet the modest 30% fuel-efficiency and GHG emissions-reduction targets imposed by the "clean car" regulations cited in Chapter I. Initial or interim vehicle design improvements that still rely on fossil fuel combustion cannot come close to eliminating persistent residual GHG discharges from future vehicle models. GM evidently was not "going all-out" to develop GHG-free electrical or hydrogen fuel-cell vehicles, and instead was making comparatively small investments in GHG-free or very-low-GHG technologies because the applicable regulatory laws only imposed limited requirements for GHG emissions reductions.

If we spend many billions of dollars on interim technology innovations, it could create an aura of legitimacy for new vehicles that in the aggregate will still discharge large quantities of persistent residual GHGs every year. When the auto manufacturers adopt first-step or interim technological changes at relatively high expense, they will be more likely to resist replacing those interim designs with carbon-free vehicles or very-low-carbon models; and therefore they are likely to challenge the imposition of stricter GHG standards in regulatory proceedings and obstructive court cases.[370] GM apparently was already planning for this kind of resistance by claiming that their customers want "affordable" vehicles,[371] before they were overtaken by the unexpected international financial crisis of 2008. Since then, GM has been getting plenty of greenwash publicity about the Volt, while they plan on selling millions more conventional GHG-polluting vehicles.

When initial or interim technological improvements reach a point of diminishing returns in design efficiencies — there is an intrinsic limit on how "clean" an internal-combustion engine using fossil fuels or biofuels can be made — it is not at all clear that the cumulative costs of interim model changes would be less than the costs of an immediate transition to carbon-free vehicles during the next two decades. Billion-dollar investments in hybrid auto engines, as one illustration, would still leave future motor vehicles dependent on harmful fossil fuel combustion and would retain little market value when polluting nations must eventually convert their automotive transportation systems to GHG-free methods.

Some transportation experts have emphasized that with increasing demands for greater mobility and increasing wealth in developing nations, we need to plan for a future doubling of global motor vehicle numbers to at least a total of around two billion cars.[372] This is a daunting prospect from a climate change perspective. If we rely on interim design measures that continue to use fossil fuels and emit residual GHGs, such as hybrid and diesel engines, these two billion vehicles will become a major source of GHG pollution that will make climate change appreciably worse and could only be eliminated through the adoption of GHG-free vehicle technologies. If we stay on the current course, we will be paying many billions of dollars to produce billions of GHG-polluting cars that will continue increasing climate change hazards.

Interim political mandates to reduce GHG pollution levels by only a modest percentage rate in the next few decades will lead to the selection of corresponding interim technologies that can meet modest regulatory targets while continuing to discharge large amounts of persistent residual greenhouse gases. Under the California and federal vehicle regulations, for example, new vehicle buyers would have to pay the costs of "first step" improvements to meet the recent fuel-efficiency standards, of probable "second step" improvements to meet stronger emissions-reduction targets in the subsequent decades,[373] and of the "third step" conversion to GHG-free technologies that will be essential in the long run. This is not self-evidently a cost-effective approach and it may be especially wasteful because, as shown in the discussion of Table 1 in Chapter II, the choice to rely on hybrid vehicles would not yield any significant climate benefits and instead would only *reduce the increases* in the atmospheric GHG concentration.

The same kind of contrast can be drawn in fossil fuel energy industries among initial or interim investments for better end-of-the-pipe GHG controls,[374] coal liquefaction or gasification projects,[375] and power plant conversion from coal to natural gas as fuel,[376] as well as Carbon Capture and Storage (CCS) technologies that would inject large quantities of CO_2 underground instead of allowing the persistent gases to reach the atmosphere.[377] Once we develop first-step technologies, such as "clean coal" gasification, they will gain substantial political support from coal-producing states, coal companies, and affected labor unions[378]; and it will become more difficult to abandon these expensive initial or interim steps that can never achieve sufficient GHG emissions reductions.

Even if fossil fuel industry innovations can attain fairly significant emissions reductions, they would still use carbon-based production processes that discharge residual GHG emissions every year, increasing the atmospheric GHG concentration. As one widely-criticized illustration, the World Bank is funding an Indian project to build a large "supercritical" coal-burning power plant — this high-temperature combustion process will cut GHG pollution by about 10% per unit of energy output in comparison to ordinary "subcritical" coal-based power plants.[379] And yet the supercritical coal-combustion plant would discharge more than 20 million tons of GHGs per year for several decades, an irrevocable "reducing the increases" GHG pollution control choice, when India could instead be moving further and faster toward increased solar power and wind power renewable energy or other GHG-free technologies.[380] How likely is it that India will give up the large investments it has made in supercritical coal-fired power plants when renewable energy replacement technologies become available? Billions of dollars expended on first-step or interim pollution controls, including economic incentive programs, and other "reducing the increases" measures for cutting GHG emissions from fossil fuel energy industries will not be available to pay for replacing fossil fuel-combustion power plants with carbon-free energy technologies.

The businesses forced to absorb high emissions-reduction costs to implement initial or interim measures will make strong equity claims that they should not have to *pay twice* when we later attempt to make them eliminate all of their "dirty" GHG-based polluting technologies. It is true that the transition to a GHG-free economy in the aggregate may take several decades, but it will also take decades for interim or first-step GHG emissions-reduction programs to impose strict controls on existing carbon-based energy sources, such as "clean coal" projects. Clean replacement technologies and graduated emissions-reduction programs will use entirely different pollution-control methods; and any pollution control program that requires large investments in GHG emissions reductions from fossil fuel-burning power plants will slow the diffusion of clean, carbon-free energy production technologies.

When we do begin the expensive process of converting to a carbon-free economy, there often will be little lasting value from the many billions of dollars expended on interim emissions-reduction measures that must be phased out along with the associated carbon-combustion energy technologies. And there will be strong political opposition from the fossil fuel energy companies, fossil fuel-producing countries or states, and other traditional fossil fuel energy constituencies that have been required to make major interim investments to reduce their GHG pollution.

Under first-step or interim emissions-reduction mandates, the American public will be asked to expend a cumulative sum of hundreds of billions of dollars on GHG pollution controls and to accept many lifestyle sacrifices in return for no tangible climate change benefits. All of the consensus GHG emissions-reduction programs involve "*reducing the increases*" outcomes that cannot improve the atmospheric GHG concentration and hence cannot provide meaningful relief from global warming and climate change. The absence of climate change benefits commensurate with high initial or interim emissions-reduction expenditures raises the unpleasant prospect that these failing measures will weaken short-term and long-term public support for GHG-free replacement technologies.[381]

The US government could mandate the rapid adoption of particular technologies to reduce GHG pollution in designated contexts or sectors, as Congress did by requiring motor vehicle fleets to burn a percentage of ethanol-based biofuels produced largely from American corn crops.[382] However, this agricultural-subsidy choice turned out to be a mistake for numerous reasons,[383] and it calls into question the desirability of regulations that impose specific pollution control technologies rather than adopting an emissions cap that would allow polluters to choose their own means for progressively decreasing GHG discharges. The biofuels legislation is another example of the "*we have to do something now*" mentality without a careful assessment of the potential downside of this risky innovation.[384] Despite the urgency resulting from climate change "tipping points" and the persistence of greenhouse gases, we cannot afford to waste billions of dollars and years of irreplaceable time on supposed remedial measures that in practice would prove more harmful than beneficial.

I am not contending that all initial and interim mitigation measures would be ineffective or wasteful.[385] Programs that encourage interim efforts to increase electric energy conservation could be extremely useful because they would reduce the aggregate amount of energy production required during the transition from fossil fuel energy generation to GHG-free alternative sources. The critical problem is to make sure that initial or interim measures will really accomplish what we want them to do at an acceptable price. This could be true for energy conservation programs and totally untrue for back-loaded GHG emissions-reduction programs, as I have argued. All environmentalists in the climate-policy field want to make rapid progress toward curtailing climate change risks; but supporting ineffective, expensive mitigation programs because we have to do *something now*, rather than because they have a good chance of actually succeeding, is a foolish climate-policy mistake.

From a political perspective, many climate-policy experts claim that we should "get the ball rolling" by initially accepting weak legislative and regulatory initiatives in the hope that these programs will grow stronger, ideally much stronger, over time.[386] Yet, based on my

35-year career as an environmental law professor, I see little empirical support for the presumption that our environmental laws will normally become stronger or better enforced with the passage of time. This happy vision might apply to the Clean Air Act, but the Clean Water Act,[387] Superfund Act (CERCLA),[388] Endangered Species Act,[389] National Environmental Policy Act, Toxic Substances Control Act, and other US environmental regulatory programs are no stronger in practice or in language now than they were during the environmental law formative years of 1969 to 1981 before the Reagan Administration was inaugurated. In practice, adopting weak mitigation measures accompanied by self-congratulatory publicity is likely to delay or defuse crucial political pressures for enacting more ambitious environmental regulatory programs.

The arena of conflicting climate policies is so complex and controversial that I do not see any rational grounds for presuming that unduly complicated, partly-corrupt, wholly ineffective mitigation proposals, such as those in the Waxman–Markey and Kerry–Lieberman Bills, would evolve into successful environmental regulatory programs able to make strong contributions toward overcoming global climate change. If anything, the opposite assumption is more probable because there are so many opposing forces pulling in divergent directions.

Many of my environmental law colleagues, who have become similarly frustrated at the lack of congressional action on climate change issues crucial for the preservation of the human race and nature, have tried to fill the climate-policy vacuum by instigating state judicial cases, state regulatory programs, municipal land-use standards, efficient business commitments, and positive consumer behavioral changes. Despite their benign intentions, this patchwork-quilt of decentralized mitigation activities has not produced any significant benefits, while it has been consuming much of the advocates' time, money, and personal energies. Proponents of effective climate change preventive actions must learn to understand the difference between promoting significant climate change progress and the much more common result of only *reducing the increases* in the atmospheric greenhouse effect and related climate change harms.

Under current adverse circumstances, no one can offer a comprehensive political solution for climate change problems that will harmonize all conflicting positions, satisfy all opposing interest groups, tie up all loose strings, and promote worldwide cooperation enabling constant mitigation successes. What I can do is present a thoughtful mitigation strategy with a reasonable chance of significant successes, not a wishful-thinking fantasy, that could be adopted when more fertile conditions for climate-policy legislation and negotiations begin to emerge. Perhaps we will need to wait until economic prosperity returns to the developed nations; or perhaps GHG regulatory efforts will depend upon negative publicity and fears stemming from catastrophes that could be caused by climate change, such as another Hurricane Katrina, or devastating Pakistan floods, Russian heatwaves, California wildfires, East African and Southern China droughts, and similar recent disasters. We cannot predict when favorable political and economic conditions will prevail again, but the practice of supporting inadequate mitigation measures under undesirable political conditions because *we have to do something now* is bound to be a climate-policy mistake with irreversible repercussions. We cannot afford to do the *wrong things* now.

The Clean Technology Commission and Development Fund

The central theme in Chapter V is that successful climate change mitigation efforts will require creating several institutions with overlapping responsibilities to meet pollution control and sustainable development objectives under diverse social, economic, and ecological conditions.[390] The development and deployment of GHG-free replacement technologies is the primary goal underlying my proposals, or one might say the lynchpin of my plan, but at least four institutions will be needed to implement this mitigation approach: (1) a government-sponsored independent commission composed of leading scientists, engineers, and economists, possibly with a few agency officials. This Clean Technology Commission would choose which replacement technologies and decarbonization methods should receive subsidies or other kinds of development and marketing support. The Commission would generally promote technological research and development activities, clean technology dissemination, international technology transfers, and greater fiscal support for innovative GHG-free technologies; (2) a progressively increasing carbon tax intended to raise the money needed to pay for the replacement technology development and deployment functions, and also to provide additional incentives for GHG pollution sources to develop their own clean technologies and to reduce their GHG emissions; (3) a direct-regulation program imposing standards for the best available GHG-free or low-GHG technologies in the sectors that generate the largest amounts of GHG discharges, on the mitigation rationale that market-based economic incentive programs alone will not function quickly enough on a large-enough scale to achieve sufficient GHG pollution control; and (4) an inter-agency task force responsible for national or international emissions monitoring and disclosure requirements applicable to all large and medium GHG dischargers. Many climate experts have advocated one or another of these institutional approaches as their main mechanism for overcoming climate change problems, but this chapter contends that we will need all of these overlapping institutions working in a coordinated manner to reinforce each other and improve the chances of successful mitigation outcomes.

The Clean Technology Commission would evaluate the trade-offs among numerous competing GHG-free replacement technologies that may be worth distributing in their present form or improving through additional research and development efforts. The Commission would also oversee the operation of a Technology Development and Dissemination Fund (TDD Fund) to provide sufficient financial assistance for worthwhile clean-technology projects, especially promising innovations developed by private companies with limited fiscal resources.

Some experts claim that existing technologies are sufficient to overcome climate change problems if they are widely adopted, while other experts contend that adequate GHG-free technologies have not yet been demonstrated in many GHG pollution contexts.[391] Rather than asking government officials to guess about which potential

GHG-free replacement technologies merit significant support, the Commission would rely on the composite technology assessments and risk assessments of distinguished members chosen primarily from the National Academy of Sciences and the National Academy of Engineering. Several federal agencies, including the Department of Energy, EPA, NASA, NOAA, and the Interior Department, are already sponsoring a wide array of projects related to climate change risks and responses. Agency officials would likely serve as advisors to the Commission with various responsibilities. These liaison functions would include sharing information on current agency projects and trying to reduce duplication of efforts, informing the agencies about new technological developments and dissemination priorities sponsored by the Commission, helping the Commission staff draw on agency expertise and resources, helping the Commission obtain sufficient funding, and providing guidance on relevant political and legal issues.

There is no question that politics, money, and private interest-group lobbying will influence the technological and fiscal choices of the Commission to some degree, as these forces have affected or infected all dimensions of American political and economic life. However, the Commission should strive to rise above common political machinations through the informed recommendations of experienced scientists, engineers, and economists who have been assessing GHG-free technologies and related climate change issues for a considerable time.

Many GHG-free technologies, including renewable energy alternatives, already exist but may not be in mature forms that can support widespread diffusion at affordable costs. An incomplete list would include at least half-a-dozen different types of solar energy processes, wind turbines, wave and tidal power generators, geothermal energy, increased hydropower generation, nuclear energy, hydrogen fuel cells, plasma gasification (a thermal chemical process to convert garbage and industrial wastes into "clean" biofuels), methane combustion from waste disposal sites and feed lots, and diverse biofuels made from nearly every biological material. In addition to various energy production technologies, Carbon Capture and

Storage methods could be useful by allowing the continued use of fossil fuels in suitable geographic areas without the pernicious atmospheric greenhouse effects; and carbon sequestration resulting from reforestation or the slowing of deforestation could also be quite valuable.

We cannot yet choose which potential clean technologies will become the most effective substitutes for GHG-polluting methods, and it is likely that different GHG-free replacement technologies will prove more or less desirable in different places under different conditions. Someone will have to make the critical GHG-free technology research and development choices and parallel funding choices. Although there are no perfect institutions, I prefer reliance on the Clean Technology Commission's technological and economic decisions to the alternatives of purely political judgments or market-forces results that are shaped by profitability expectations more than by human and environmental welfare considerations.

The central feature in comparing an array of complicated and often competing GHG-free technology candidates is that the Commission would be primarily composed of scientific and engineering experts who can make reasonably well-informed, not perfect, judgments on the relevant technical issues. Then the Commission can draw upon the Development Fund to provide financial support for whichever clean technologies they conclude will probably be the most effective and affordable.

Climate mitigation proponents should not be wasting their efforts and limited resources on inherently inadequate GHG emissions-reduction programs, and they should not be distorting the allocation of pollution control resources to ensure the profitability of fossil fuel industries with many lobbyists and politically powerful clients. Some institution will have to make many difficult mitigation choices and trade-offs, and the vital question is whether a relatively impartial commission of scientists, engineers, and economists is likely to make better impartial judgments about alternative clean technology options than the choices made by purely political regimes or market forces dominated by fossil fuel companies. It will not be possible to exclude all political and market-system influences from the required transformational process, but relying primarily on the Clean Technology Commission to offer informed, transparent reasons for their judgments appears to be a more reliable and desirable process than depending primarily on the choices of political or market institutions.

As one illustration, Dr. Stewart C. Prager, the director of the Princeton Plasma Physics Lab, recently published an Op-Ed piece in the *New York Times* contending that unlimited fusion power fueled by deuterium and tritium drawn from seawater could become a dominant energy source for almost all nations within a 20-year period.[392] He acknowledged that this innovative energy-technology approach, which would be entirely GHG-free, will be moderately expensive during the transition from research and development efforts to commercial applications, but he emphasized the almost incalculable benefits of this inexhaustible source of clean energy. Dr. Prager then complained that the US is being left behind by competing high-technology nations, including China, Japan, South Korea, and Russia, because of the US reluctance to make the essential investments.[393]

Can you imagine a comparatively low-cost, nearly unlimited, entirely clean replacement technology that could supplant the major fossil fuel producers and "dirty" energy producers in a relatively short interval? Can you imagine an alternative energy technology that does not require importing many billions of dollars of oil from foreign countries, that does not require dangerous drilling and mining operations, that does not require expensive pipelines with associated storage facilities, and that would be completely GHG-free in operation? Understandably, this marvelous prospect seems too good to be true, and the fusion process is likely to have distinctive costs and risks of its own.

Who should determine if this replacement technology is worth pursuing? There seems little doubt that the Clean Technology Commission would be a better qualified and more reliable institution to evaluate the validity of Dr. Prager's fusion proposal and to subsidize the development of this clean alternative technology if the Commission experts find this is a credible prospect. No institutional mandate is foolproof, but the thought of leaving this critical decision to short-sighted politicians or energy markets dominated by trillion-dollar fossil fuel companies is more than a little chilling. At this present time, according to Dr. Prager, these self-interested, short-sighted decisionmakers are wasting a major opportunity for the US to take the lead in developing this clean fusion technology.

This description of a potentially effective, reasonably affordable source of GHG-free energy that could function anywhere that has access to seawater is exactly the kind of innovative technology on the horizon that should be carefully evaluated and actively supported if it appears to be a viable replacement for harmful GHG-polluting energy sources. If Dr. Prager is right, we could eliminate the worst categories of GHG polluters in far less time and possibly at far less cost than any climate expert has previously predicted. This is an exciting opportunity, and also an inexcusable tragedy that the American government is not pursuing clean fusion technology potential in as thorough and rapid a manner as possible.

As Professor Prager emphasized, we have been falling behind other scientifically proficient nations as a direct consequence of our ineffective political and economic institutions. I am not arguing that reliance on a commission of senior scientists, engineers, and economists to select meritorious technological improvements is a perfect institutional process, but I do contend that it is far more promising than what we have now. We can do better and must do better, which will only be possible if we recognize the weaknesses of our present policy-making institutions and create better ways to make better decisions.

Because we usually cannot predict which GHG-free technologies will prove most effective over time in different contexts, the Commission will need to sponsor numerous design, production, and pilot projects that would enable various clean technologies to compete against each other. However, it is crucial that these GHG-free technologies must not be expected to compete against GHG-based technologies, such as the present energy, transportation, and manufacturing processes that rely on fossil fuel combustion. The fossil fuel producers possess vast financial resources and cannot be allowed to under-price or under-bid nascent clean energy technologies, as they did in the aftermath of the oil crisis of 1973. We must not forget that long-established fossil fuel-based technologies have been the leading anthropocentric cause of global climate change, and we cannot allow the producers of fossil fuels to undermine efforts to create safer GHG-free replacement technologies.

The main functions of the Clean Technology Commission would be to assess, compare, sponsor, and subsidize an array of potential GHG-free technologies while promoting greater market penetration and insulating clean replacement technologies from the commonly distorted market prices and frequently corrupt practices associated with established GHG-based market forces. Alternative GHG-free replacement technologies probably cannot become dominant energy sources without substantial research, development, and marketing subsidies as well as various other kinds of government support.

It is true that government subsidies may be misused to promote the wrong things — this is exactly what has happened with the huge fossil fuel subsidies the US government has been handing out for more than half a century. Proponents of clean energy technologies will have to avoid the misuse of subsidy funds for foolish or selfish purposes, which is an inevitable risk of dealing with our often-frustrating political system. Yet, government leadership and funding for climate change programs is not impossible to imagine, and without substantial government support we probably must give up the game, set, and match.

Global warming and climate change were directly caused by the insatiable free-market pursuit of greater profits resulting from economic growth and constantly increasing consumption. Nevertheless, free-market ideologues are bound to contend that government decisions will always be even worse than imperfect market-system choices. Yet, coordinated collective action mandated by government is the generally-accepted means in conventional economic theory to overcome public-goods and free-rider problems.[394]

In response to "free market" advocates who believe that government can do no good, I would point to the Eisenhower Interstate Highway System, the Publicly-Owned Treatment Works construction program under the Clean Water Act, the Apollo program, and the

Marshall Plan as instances in which the American government has undertaken multi-billion-dollar projects with appreciable success. The potential harms from global climate change are certainly as pressing as the public needs that inspired these ambitious government programs, and the need for prompt collective action on climate change perils is equally or more urgent.

The Clean Technology Commission cannot perform the various tasks identified above without adequate funding, which explains the requirement for the Technology Development and Dissemination Fund (TDD Fund). In the past decade, many international and national Funds have been created to offer financial assistance for climate change mitigation and adaptation programs.[395] For example, as of 2008 the World Bank was participating in the financing of a dozen climate-related Funds, including one with about $5 billion to help developing countries adapt to climate change hardships and another large Fund to help these poor states achieve cleaner energy production.[396] The latter type of grants were sharply criticized by environmental groups because the World Bank has been willing to support "clean coal" projects and other fossil fuel energy technologies that will produce large volumes of residual greenhouse gases for several decades.[397]

None of the climate change Funds I have located has been focused exclusively on the development of clean GHG-free replacement technologies with the extensive financial resources that should be available to the TDD Fund under my plan. A very substantial amount of financial support will be required to pay for comparative evaluations of dozens of GHG-free candidate technologies and to promote the design, development, marketing, dissemination, and protective nurturing of feasible clean GHG-free technologies that can replace GHG-polluting methods in various contexts.[398]

Expending a few billion dollars here and a few billion dollars there on many different mitigation missions and institutional preferences is unlikely to produce significant progress anywhere. We should put the great majority of our climate change funding "eggs" into a GHG-free replacement technology "basket" because only that approach can meet the goals of both the developed and developing nations while stabilizing and then reducing the atmospheric GHG concentration. The world's nations need to adopt clean replacement technologies that do not create persistent residual GHG discharges, which is the opposite of the consensus emissions-reduction approaches that have been dominant in the past two decades.

A Progressively Increasing Carbon Tax

The main purposes of this carbon and carbon-equivalent GHG tax are to provide ample funding for the activities of the Commission and TDD Fund, and also to create an additional deterrent incentive that would help persuade polluting sources to reduce their GHG discharges. The former function is crucial for the widespread development and deployment of clean replacement technologies, while the latter function attempts to reduce GHG discharges by gradually increasing their economic costs.

The carbon tax would raise the price of discharging GHG emissions by an amount chosen on the basis of whatever price level could provide the needed revenues and could induce the desired emissions reductions in a realistic manner. The carbon tax price would not actually reflect the full social costs and benefits of the activities that produce GHG pollution and the resulting climate change harms. "Putting a price on carbon" means raising the cost of the polluting operations by a given tax amount, not attempting to reach the speculative market price that theoretically would prevail if all associated costs and benefits, including all externalized harms, were imposed on the polluting activities. In effect, "putting a price on carbon" is a government function that would be based on a typically crude estimate of how much pollution reduction could be achieved by using a chosen tax level. In realistic terms, "putting a price on carbon" has very little to do with a "free market" system except that after the carbon tax is imposed by government, individual firms can choose whether to pay the tax or to find some way to reduce their GHG discharges.

Now that cap-and-trade systems appear to have lost their political credibility in the US Congress, carbon taxes are becoming the "in thing" in various contexts. The influential Yale economist, William Nordhaus, recently proposed that a carbon tax should be imposed to help reduce the federal government's massive deficit.[399] He explained that the carbon tax "is virtually the only tax under consideration that will increase economic efficiency because it reduces the output of an undesirable activity (carbon dioxide emissions). Every other tax that is under discussion will reduce economic efficiency."[400] Professor Nordhaus also noted that the carbon tax would increase the price of carbon emissions, a "necessary condition for slowing climate change."[401]

Professor Nordhaus asserted that a "carbon tax can buttress or replace many inefficient regulatory initiatives and will thereby provide yet another improvement in economic efficiency."[402] I agree with him that a carbon tax can "buttress" imperfect regulatory programs, which is part of my plan, but to the best of my knowledge Professor Nordhaus has not discussed whether regulatory systems may be equally necessary to "buttress" imperfect carbon taxes. If fossil fuel producers are able to use their great wealth to pay the carbon tax without changing their GHG-pollution practices, regulation may be essential to counter-balance the mitigation weaknesses of carbon tax incentive schemes. From a climate mitigation perspective, we cannot allow fossil fuel sources to continue contaminating the atmosphere regardless of whether they are willing to pay whatever carbon tax the government imposes.

Robert B. Reich, an economist who was the Secretary of Labor during the Clinton Administration, recently published a book blaming the increasing income disparity in the US for many of the structural problems in our economy, and he suggested that the government could use the revenues from a carbon tax to reduce income and wealth inequalities.[403] He was simply treating the carbon tax as a prospective pot of money available for uses that have nothing to do with reducing climate change dangers.

Dr. Hansen's fee-and-dividend system is the same as a carbon tax and rebate scheme that would return all carbon tax revenues to American consumers and their families.[404] Before he devised this idealized system, Dr. Hansen devoted more than 20 years to explaining how harmful the emissions from coal-fired power plants are and he proposed eliminating all coal-fired facilities as rapidly as possible.[405] And yet, as discussed previously, Dr. Hansen did not explain in any realistic detail how the modest carbon fee (tax) he advocates could transform the world's economy and energy systems, and would be able to overcome the dominant market power of coal-fired power plants and other fossil fuel industries.

From the perspective of climate change mitigation, the fundamental problem with these diverse carbon tax proposals is that they do not provide funding designated for the development of clean replacement technologies that would reduce the atmospheric GHG concentration and climate change harms. Dr. Nordhaus only refers to the climate benefit of putting a price on carbon emissions, not to the need for financing the development of less harmful technologies and methods. Dr. Reich views the proceeds from carbon taxes in the same manner as any other "sin tax." And Dr. Hansen's impassioned, scientifically valid criticisms of coal-fired power plants have not been able to forestall the reality that China and India are rapidly increasing the number of their coal-burning power generators and even the US has approved plans for dozens of new coal-fired power plants.[406]

Climate protection advocates have failed to create and publicize effective, affordable, convenient GHG-free replacement technologies that would be able to expand worldwide energy production without destroying our climate. In the absence of affordable clean-energy technologies, most nations will not abandon the long-established, least-costly, most familiar, "dirty" method for producing more energy no matter how many criticisms are directed at King Coal. The many consensus emissions-reduction programs of the past decade were fatally flawed because they focused exclusively on cutting GHGs and devoted little or no attention to the need for clean replacement technologies that could provide the increasing energy and industrial capacities that most developing countries consider their highest priority.

The weakness of Dr. Hansen's anti-coal position, as scientifically justified as it may be, is that he did not present a compelling case for the creation of alternative clean-energy sources. We must develop adequate clean-energy alternatives *before* asking the nations of the world to give up proven fossil fuel technologies that will promote desired economic development despite causing increased global warming. If climate mitigation advocates, such as Dr. Hansen, cannot provide feasible GHG-free technology alternatives, the request to abandon coal-burning power plants and other "dirty" GHG pollution sources is an untenable appeal for many poor countries and impoverished peoples to *sacrifice* the economic and social development they want most.

Instead of telling billions of people in developed and developing nations to reduce this activity and cut out that activity, and to give up many desired amenities and activities, as environmentalists have often advocated, a more realistic strategy must be based on the dissemination of affordable GHG-free technological alternatives that people find satisfactory because they will still be able to do what they want to do while they are helping to remedy climate change hazards. The mission of the TDD Fund to support the transfer of clean replacement technologies to developing nations is critically important because climate protection campaigns based on pleas for widespread public sacrifices cannot succeed in comparison with the intensely-felt economic and social welfare aspirations in nearly all developing countries.

If we do not use the revenues from the carbon tax to support the functions necessary to design and disseminate affordable GHG-free replacement technologies, where is sufficient funding going to come from? No matter how much we complain about fossil fuels and the long-term damage they are causing, we cannot overcome them without viable alternatives that will probably be quite expensive at the beginning of the technology-replacement process. As a result of the need for substantial financial support, I cannot approve the imposition of any carbon tax or cap-and-trade proposal that fails to allocate a major portion of the net revenues for the purpose of funding the deployment of GHG-free alternative technologies.

The carbon tax revenues may also be needed to support redistributive activities among people who are likely to be disadvantaged by the transition from fossil fuels to clean replacement technologies. Fifty years ago, the Interstate Highway System and Publicly-Owned

Treatment Works construction programs, although they provided large net social benefits, both contained redistributive provisions to soften the costs and dislocation burdens imposed on people in relatively disadvantaged areas. The sponsors of GHG-free replacement-technology programs, GHG pollution control efforts, and climate change adaptation programs will similarly need to consider the political dimensions of national and international redistributive effects. There will be some perceived "losers" no matter what is done, even if we are able to create effective mitigation programs that can overcome global climate change. The question is how much political clout the "losers" will have, and whether climate mitigation advocates must find ways to mollify them. Redistributive or remedial efforts will probably require reliable funding that must come from somewhere.

Federal or state governments in the US will almost certainly need to offer social dislocation and retraining programs in regions that now benefit from fossil fuel production as a means to gain sufficiently broad political support for the adoption of alternative clean-technology initiatives. In the "real world" of divergent human interests and priorities, many people will not vote, or voluntarily pay, to support GHG pollution control efforts that threaten their jobs and incomes, their lifestyles, or other significant losses. We must also take into account the cognitive psychology finding that many people would rather avoid a loss of what they already have than pursue a comparable gain.[407] Effective programs to combat climate change are certainly going to result in losses for some people, communities, and businesses; and we will likely need to provide some form of dislocation assistance to help overcome the resulting political hostility. Again, it is unclear where the essential funding will come from if we do not create a carbon tax or other explicit funds-raising regime that can produce the necessary revenues.

When I circulated an earlier version of this chapter in the hope of obtaining useful criticisms, I received several comments from a law professor who shall remain anonymous because I did not indicate that his reply might be published. He wrote: "I am unconvinced by the comparisons of a cap-and-trade regime that is full of holes and exceptions to a carbon tax regime that is pure. The same political forces that contaminate a cap-and-trade system will work equally hard, and with equal effectiveness, at contaminating a carbon tax regime." I was rather amused to be accused of the same kind of "asymmetric idealism" that I have sharply criticized in others, but I do not agree that I have adopted an unrealistic analysis in assessing the strengths and weaknesses of carbon taxes.

The machinations of self-serving politicians, private-interest groups, and a flock of eager lobbyists may produce an equally distorted tax scheme at its inception (in comparison to cap-and-trade systems). One need only look at the many loopholes and special-interest benefits in the regular income tax code to see that this is true. However, after the tax scheme is established, it will be considerably simpler to administer a carbon tax applicable to all substantial GHG discharges than a cap-and-trade system in which carbon-market speculators and market-making profiteers have strong ongoing incentives to manipulate the system as a way to increase their personal returns.

All carbon tax revenues would be collected by the government, and under my plan most of this money should be allocated to the TDD Fund. In contrast, under cap-and-trade schemes in the Waxman–Markey and Kerry–Lieberman Bills, the government would have given away most of the GHG allowances to grandfathered pollution sources and those "dirty" firms would collect most of the revenues from the cap-and-trade system. A cap-and-trade mechanism does not have to be designed this way because the government could obtain large revenues by auctioning most of the allowances rather than giving them away. Yet, the legislative proposals would have enriched many of the worst GHG-polluting companies instead of using the funds obtained to develop clean GHG-free technologies. If one considers how the cap-and-trade systems would actually be administered under the recent US congressional bills, there is a world of difference between these polluter-friendly, give-away cap-and-trade allowances systems and a carbon tax with the revenues primarily devoted to creating and disseminating better GHG-free technologies. However, this legal critic was certainly correct that the revenues from a carbon tax could be misallocated for purposes that do not strengthen climate change mitigation.

Another important distinction is based on the types of results that these competing economic incentive models would likely yield. The Interstate Highway System was financed by dedicated gasoline taxes that did not evoke much public opposition because people could see the construction of new highways and the practical benefits arising from improved transportation. If my proposals for the TDD Fund applications of carbon tax revenues are adopted, people will similarly be able to see the increasing market penetration of clean GHG-free replacement technologies as they are gradually deployed on a widespread basis. In contrast, the legislatively-proposed cap-and-trade systems are likely to lose their degree of public tolerance much sooner because these programs will lead to "reducing the increases" outcomes that will not yield any meaningful climate change benefits. Taxpayers will hear about numerous trades of GHG allowances, which most people will not be able to understand, and they will hear about the many consultants and speculators who would populate the GHG allowance-trading field, but concerned people will not be able to identify any discernable climate change benefits from the cap-and-trade process. This is true because there will be no tangible benefits, as explained in Chapters II and III.

The combination recommended here of progressively increasing carbon taxes and direct pollution-control regulations in high-GHG industrial sectors would create a continuing incentive for firms to reduce their pollution even when an established business, such as an oil company or electric utility, could readily afford to buy the required number of pollution allowances under a cap-and-trade program[408] or to pay the mandated carbon taxes. A major reason why a direct regulation component is included in this multi-institutional plan is to ensure that wealthy GHG dischargers cannot subvert the mitigation program by continuing to put out large amounts of pollution because they are able to buy enough GHG cap-and-trade allowances or to pay the carbon taxes imposed. This is a good illustration of how relying on different but compatible institutions with overlapping responsibilities can help prevent the circumvention or distortion of a regulatory regime by GHG polluters that do not want to clean up their operations.

"Technology-Based" Regulations in High-Pollution Sectors

Until the late 1960s, pollution control in the US was regarded mainly as a responsibility of the states, which were supposed to determine how harmful specific levels of specific pollutants are, and what levels of pollutant discharges should be chosen as acceptably safe. Very little pollution control progress was achieved under this "harm-based" regulatory approach. In the 1970 Clean Air Act and even more in the 1972 Clean Water Act, the US Congress increasingly federalized pollution control responsibilities and shifted the primary approach from harm-based regulation to "technology-based" regulation.[409] Under the latter approach, EPA was required to determine the best available technologically and economically feasible pollution control methods for each class or category of industrial polluter.

Under this regulatory approach, EPA was not required to make scientifically reliable assessments of how dangerous varying discharge levels of each regulated pollutant would be, or what degree of harm should be deemed *acceptable* (or *unacceptable*). Instead, the agency was required to determine which pollution control methods were the best technologically and economically feasible ones. During the next few decades up to the present, technology-based regulation has overwhelmingly been adopted in preference to harm-based regulations and economic-incentives-based regulations in the contexts of toxic water pollutants and hazardous air pollutants because of technology-based regulation's greater implementation practicality.[410]

This is a simplified overview of a complicated, time-consuming, costly, often frustrating regulatory process with one compelling rationale: The vast amount of data on pollution control technologies and feasible economic impacts required to implement technology-based standards has proven to be far more attainable and defensible, although expensive, than the acquisition and validation of scientifically reliable risk-assessment data needed to set harm-based regulations, which has proven to be virtually impossible for many pollutants.[411]

As a result of the shift from harm-based regulation to technology-based pollution-control regulation, the US EPA has accumulated extensive experience in setting different technology-based standards for different industries that discharge different mixes of pollutants, which in practice has required EPA to issue thousands of detailed rules and regulations applicable to hundreds of distinctive classes and categories of polluting industries. In contrast, my proposals would impose a less comprehensive version of "best available" technology-based standards targeting only the GHG-polluting industries discharging a substantial portion of total US GHG emissions. These regulated industries would include fossil fuel producers and power plants, transportation sectors including oil refineries and vehicle manufacturers, cement factories, and perhaps building construction and feedlot operations facilities. The lesser-polluting industrial dischargers in other sectors would be subject to the carbon tax but not to technology-based standards, unless the EPA determined in a specific industrial context that a particular polluter or a class of polluters was contributing substantially to excessive GHG emissions.

This direct regulation plan is somewhat similar to, but also crucially different in design from, the various regulatory measures that the US EPA has chosen in response to its regulatory finding that GHGs represent continuing threats to human health and welfare.[412] With regard to technology-based standards, EPA has chosen to begin regulating refineries and fossil fuel-burning power plants, "sources that make up nearly 40 percent of the nation's greenhouse gas emissions."[413] In a critical mistake, EPA plans to issue technology-based restrictions for GHGs following the New Source Performance Standards (NSPS) pattern applicable to other air pollutants, in which only new plants or substantial renovations of existing plants will be covered by the EPA regulations. The great majority of GHG emissions from existing refineries and power plants will supposedly be regulated by the states, or by no one.[414] In my view, we cannot afford to consign existing coal-fired power plants and other large GHG-polluting sources to this kind of capricious decentralized regulation efforts.

Reliance on state GHG pollution control is a foolish climate-policy mistake because numerous American states are now teetering on the equivalent edge of bankruptcy and others depend on jobs and revenues from fossil fuel production industries that they are very unlikely to risk threatening. Perhaps the bifurcated "new plant" and "existing plant" regulatory strategy was warranted for other air pollutants with fewer global effects and localized economic ramifications, but putting the majority of the GHG emissions-reduction responsibility on states that typically have inadequate administrative funding and capacities, and often divergent political perspectives, is bound to produce counter-productive inconsistencies and delays. All of the reasons why the US shifted from feeble state-managed pollution-control regulation to federal pollution-control standards 40 years ago appear just as valid today with regard to greenhouse gas regulation.

For GHG sources not included in the refinery and power plant industries, EPA intends to rely mainly on a decentralized permit-negotiations process rather than on specific technology-based standards for different polluting sectors.[415] As shown in Table 4, EPA

Table 4: Industries for EPA Greenhouse Gas Permit Regulation (Based on the North American Industry Classification System)[416]

Industries
Agriculture, fishing, and hunting
Mining
Utilities (electric, natural gas, other systems)
Manufacturing (food, beverages, tobacco, textiles, leather)
Wood product, paper manufacturing
Petroleum and coal products manufacturing
Chemical manufacturing
Rubber product manufacturing
Miscellaneous chemical products
Nonmetallic mineral product manufacturing
Primary and fabricated metal manufacturing
Machinery manufacturing
Computer and electronic products manufacturing
Electrical equipment, appliance, and component manufacturing
Transportation equipment manufacturing
Furniture and related product manufacturing
Miscellaneous manufacturing
Waste management and remediation
Hospitals/Nursing and residential care facilities
Personal and laundry services
Residential/private households
Non-Residential (Commercial)

has identified a wide range — too wide in my view — of GHG sources as well as some non-industrial sources such as hospitals and residential households to be regulated mainly at the permit level, if at all.

It will take several decades, which is much too long, for EPA to implement effective technology-based emissions-reduction regulations for this number of distinctive industries with markedly different sectoral characteristics. This temporal inadequacy may be a consequence of attempting to employ technology-based regulation as the only applicable mitigation approach, although the projected timeframe is incompatible with curtailing the rapid worldwide growth of the atmospheric GHG concentration and greenhouse effect. In contrast, my recommended strategy would use four separate institutional approaches with partly overlapping responsibilities, and it would limit the "best available" technology-based regulations to only the most heavily polluting industries, such as power plant utilities, that possess a variety of feasible technological alternatives.

Under its current plans, EPA will restrict most categories of GHG pollution sources listed in Table 4 to regulatory standards imposed under the Prevention of Significant Deterioration (PSD) provisions and Title V (permitting requirements) of the Clean Air Act. As of January 2, 2011, "Step 1" of the PSD requirements, "most notably, the best available control technology (BACT) requirements, will apply to projects that increase net GHG emissions by at least 75,000 tpy [tons per year] carbon dioxide equivalent (CO_2e)."[417] Yet, in order to reduce the administrative burden for state officials and decentralized EPA offices, this GHG limit will only be imposed if the "project also significantly increases emissions of at least one non-GHG pollutant."[418] For the primarily state-run permitting program, "only existing sources with, or new sources obtaining, title V permits for non-GHG pollutants will be required to address GHGs during this first step."[419] The second step "beginning on July 1, 2011, will phase in additional large sources of GHG emissions."[420]

The EPA guidance statement for future GHG regulations mentions an undefined "third step of the phase-in that would include more sources, beginning by July 1, 2013."[421] The regulatory notice imposes "a rule that no source with emissions below 50,000 tpy CO_2e, and no [facility] modification resulting in net GHG increases of less than 50,000 tpy CO_2e, will be subject to PSD or title V permitting before at least 6 years from now, April 30, 2016. This is because we are able to conclude at the present time that the administrative burdens that would accompany permitting sources below this level will be . . . impossible to administer . . . until at least 2016."[422] EPA acknowledged this administrative limitation before the Republicans in the House of Representatives began a high-profile campaign to decrease the agency's funding.[423]

As a result of increasingly hostile political conditions, it is by no means clear that any of these complicated GHG regulations will be implemented in the near future. In June 2011, EPA chose to push back for at least several months the "release of a proposed rule on greenhouse gas emissions from power plants and other major pollution sources."[424] As one reporter noted, "the delay is a tacit admission that the regulations pose political, economic and technical challenges that cannot be addressed on the aggressive timetable that the agency set for itself early in the Obama administration."[425] He added that the "postponement is the latest step by the EPA to slow the issuing of regulations that critics say will slow economic growth, drive up energy costs and reduce employment."[426] If the Republican members of Congress who now control the House of Representatives continue to oppose climate change programs that might raise the costs of the polluting industries and their customers, there is no reason to expect that EPA will receive sufficient funding to develop, implement, publicize, and enforce the controversial regulations.[427]

In recent regulations, EPA and the National Highway Traffic Safety Administration (NHTSA) issued "a first-ever Heavy-Duty National Program to reduce greenhouse gas (GHG) emissions and improve fuel efficiency of medium- and heavy-duty vehicles, such as the largest pickup trucks and vans, semi trucks, and all types and sizes of work trucks and buses in between."[428] This emissions-reduction program is similar to the one imposed a couple of years ago on new passenger vehicles and light trucks. As explained in Chapter II, this kind of program will *reduce* the BAU level of annual GHG discharges from motor vehicles and trucks, but it will also allow the residual GHG discharges from trucks to *increase* the atmospheric GHG concentration and related climate change dangers.

One serious problem underlying this convoluted multi-part, multi-source EPA regulatory scheme, which is more complicated and has more exceptions than these quoted passages reveal, is the absence of alternative deterrent institutions, such as a carbon tax, that could create ongoing incentives for smaller-scale GHG sources to reduce their emissions. As a result of adopting only one regulatory process and institutional regime for each category of regulated industry, EPA has chosen to make many compromises between effective GHG pollution control and difficult implementation problems that in practice will reduce the effectiveness of the GHG restrictions imposed in order to decrease the administrative burdens and political conflicts from regulating many thousands of comparatively smaller sources.

In essence, EPA is trying to apply too many different technology-based approaches to fit too many distinctive categories of GHG sources because this is the main process that they have used in past decades to restrict other types of air pollutants. EPA is consequently including many exclusions and exceptions that could apply improperly to the worst GHG polluters and will certainly complicate regulatory implementation for all GHG sources. As one example, there is no reason for EPA to limit the regulation of GHG dischargers to only those polluters that are also dischargers of other regulated pollutants, except for the lower administrative costs of modifying already-existing pollution permits that were imposed on the other pollutants.

My recommendations would impose BACT technology-based standards on large and mid-sized GHG polluters in a limited number of industries that in the aggregate represent a substantial majority of the annual GHG discharges from US sources. All of these industries would have to discharge at least several percent of the cumulative US annual GHG emissions in order to be regulated under this technology-based program. Many of EPA's "Industry Groups for Greenhouse Gas Regulation" in Table 4 would not warrant technology-based regulations under my plan, and instead would be subject only to the carbon tax and mandatory disclosure provisions.

Conversely, most dischargers with GHG volumes below 50,000 tons per year of GHGs will not be regulated or incentivized at all by the EPA rules, whereas these sources would have to pay the carbon tax and submit required annual disclosure statements under my proposals. The residential and commercial sources cited in Table 4 would not be directly regulated under my scheme, and instead would be subject to higher market prices from the carbon tax imposed on GHG-polluting firms that will raise their consumer prices to reflect increased regulatory costs.

The proponents of economic-incentive systems, including cap-and-trade systems and carbon taxes, typically reject "inefficient" direct regulation programs and choose to rely entirely on the economic incentives and market forces created by their preferred programs. I do not agree that we can control GHG discharges in the worst-polluting industrial sectors by depending exclusively on economic mechanisms to overcome climate change problems during the next several decades. We must acknowledge that the fossil fuel industries control trillions of dollars of assets that will surely influence how economic-incentive schemes will shape industry behaviors, and how industry wealth will shape the performance of economic-incentive systems. We must not fail to recognize that unrestricted "free market" forces and the resulting harmful externalities have been the single largest cause of human-induced climate change; and it would be a travesty if supposedly flexible market forces and economic incentives programs are dominated by the "dirty" fossil fuel polluters that continue to exacerbate climate change dangers.

Every economist with any interest in environmental and energy issues is doubtless aware of the massive historical and current fossil fuel subsidies, widespread externalized damages, economies of scale in the production and distribution of GHG-generating goods, and various other aspects of energy, manufacturing, and transportation markets that will make it extremely difficult for GHG-free technologies to compete against the established fossil fuel combustion technologies under existing market conditions.[429] These are not "free competitive markets" in any meaningful sense. We also have no compelling market-based means to overcome the difficult collective-action and free-rider problems imposed by climate change risks. Being aware of systemic market imperfections is not the same as being able to correct them within the market system itself.

Without a comprehensive understanding of the potential benefits and perils of GHG-polluting activities that contribute to the atmospheric greenhouse effect, which is often beyond the state of scientific and economic knowledge, there is no reliable way that energy prices or GHG-pollution prices set by carbon markets could be *correctly* "corrected." Moreover, market prices depend equally on demand reflected in the willingness-to-pay choices of consumers. Yet, achieving the "correct" demand and resulting market prices for GHG-polluting activities would require billions of individuals with limited information and cognitive capacities to make innumerable "correct" purchases reflecting their preferences about energy availability, industrial processes, desired goods with associated harms, and complex climate change impacts that they would rarely understand. These uninformed consumption choices would be further degraded by high transactions costs required for reaching collective-action agreements among the enormous number of people and organizations affected by climate change and limited by free-rider problems, coordination problems, and "public goods" problems that would be very difficult, if not impossible, to resolve through decentralized market choices.[430]

Imagine a Marquess of Queensberry boxing match between the fossil fuel industries and the GHG-free renewable energy competitors. In one corner, there are well-entrenched fossil fuel producers with trillions of dollars of assets; a century of experience and widespread political connections; long-established product delivery and storage networks; extensive subsidies and economies of scale; ardent support from fossil fuel-producing nations and regions; hundreds of thousands of existing employees; a veritable army of lobbyists, lawyers, and accountants; and a great many friendly politicians who welcome campaign contributions.

In the other corner of the hypothetical ring, hundreds of relatively small and often frail businesses will be competing against each other and against the fossil fuel companies for a small slice of regional energy markets. These firms will not have any of the advantages of the fossil fuel industries except perhaps a higher capacity for innovation. These businesses may, or may not, be supported by environmentalists and climate change victims with loud voices but shallow pockets. In this analogy, the boxing-match outcome is highly likely to be a first round knock-out by the fossil fuel industries, with no realistic chance for competing GHG-free energy companies to challenge the GHG-polluting behemoths on a large scale. In light of the great imbalance in economic and political power, the societal transformations required for effective climate change mitigation progress cannot be accomplished through "free market forces" operating in a "real world" economy.

In many cases in which economically and politically powerful fossil fuel companies can pass on the cost of high carbon fees or allowances to other firms and consumers, there is no reason to assume that the GHG-polluting industries will voluntarily give up their enormous assets and will essentially eliminate themselves in response to supposed market forces. This is another "stocks and flows" problem that has not received sufficient consideration. Many economists claim that if we adopt market-forces measures and "put a price on carbon," the companies with more efficient, less harmful offerings will eventually overcome the GHG-polluting industries. But this analysis does not take into account the massively unequal trillion-dollar "stock" of wealth, assets, and political connections that the fossil fuel industries now possess compared to the clean technology firms and renewable energy processes that have hardly gained a toe-hold in our economy.

The bottom line is that effective climate mitigation efforts are likely to require GHG regulation as well as economic incentive programs. Each institutional approach will benefit from the simultaneous availability of the other approach, and each regime may be destined to fail if its inherent weaknesses are not corrected or counter-balanced by reliance on the other compatible institutional mechanisms.

From an international perspective, it would be desirable if the US, the EU, and other developed nations could cooperate in operating the TDD Fund and adopting the same GHG regulatory approach, but this seems to be an unlikely prospect. Disputes about which Fund expenditures should promote which technological developments in which countries would be difficult to resolve, and competitive tensions will arise from the ramifications of patenting GHG-free replacement technologies supported at least in part by financing from a coalition of nations with partly conflicting or partly competing business interests. At the same time, we also would need to design clean technologies suitable for technology transfers to developing nations as a means to reduce the GHG pollution levels in these countries without sacrificing their economic and social welfare opportunities. Harmonizing these often-conflicting goals will not be easy in practice. The central point, however, is that several mutually-reinforcing GHG pollution-control approaches and associated institutions may have a better chance for mitigation success than any direct regulation effort or economic incentives scheme would when acting alone.

Mandatory GHG-Pollution Disclosure Programs

Disclosure programs may be useful as primary institutions in some climate contexts and as supplemental tools in others. Following the example of the US Toxics Release Inventory (TRI) mandated by the Clean Air Act[431] and comparable EU programs, such as the "European Pollutant Release and Transfer Register,"[432] regulatory *shaming* programs can help reinforce deterrence incentives by relying on disclosure requirements and adverse publicity to embarrass major GHG sources and threaten their reputations as cooperative corporate citizens. These disclosure mechanisms could be used to supplement technology-based pollution control standards for some industrial GHG sectors and to supplement the progressive carbon tax for mitigation efforts in other sectors.

Another context in which disclosure programs could be quite useful is in attempts to change voluntary consumer behaviors. Despite many pronouncements asking consumers to reduce their "carbon footprints," virtually all programs for mandatory emissions reductions, including cap-and-trade systems and other economic incentive mechanisms, have been directed at GHG-polluting production activities from diverse industrial and transportation sectors. For example, several laws have recently been enacted that require automobile manufacturers to improve the fuel efficiency of their vehicles or to reduce their GHG emissions, but no laws require vehicle purchasers to drive fewer miles or to buy smaller cars. This differential treatment is arguably a confirmation of the presumption that we can ask many millions of consumers to lessen their GHG impacts in various ways, but it would be politically unpalatable as well as ineffective to try to force these consumers to make personal sacrifices in GHG emissions-reduction contexts.

Mandatory disclosure programs could inform consumers and businesses about the worst GHG-polluting products, services, and companies; about the relative status of competing firms in a given industry in terms of their GHG pollution volume per unit of production; about the best energy-efficient products and GHG-free product choices for consumer purchases; about behavioral and recreational practices that are especially harmful from a climate perspective; and about the existence of an increasingly wide range of clean alternative products and services. I do not believe that we could depend primarily on consumers to reduce most GHG emissions resulting from their daily behaviors, but GHG disclosure programs would enable those people who are seriously or casually concerned with climate change impacts to make appropriate purchasing decisions that can influence the competitive choices of polluting businesses whose perceived "goodwill" is at stake.

The government agencies responsible for GHG regulatory programs could choose to require more or less detailed disclosures in particular consumption settings, such as annual GHG pollution volume as a function of a company's per-unit production output and natural resources exploitation. These disclosures would improve the ability of concerned consumers to make informed choices about which goods and services they should buy, and the disclosures would also allow other consumers to limit the cognitive burdens arising from evaluating complex climate change issues. It is difficult to see how economic-incentives or market-forces regimes could succeed unless they are supported by extensive disclosures from companies that are significantly damaging the climate, because "willingness to pay" assessments cannot be effective without consumers having sufficient knowledge about what they want to pay for.

A few scholars have suggested that US regulations should require the compilation and disclosure of an "individual carbon release inventory" that would inform consumers about their personal effects on GHG pollution levels and would encourage or force them to reduce their individual carbon impacts to some degree.[433] This climate-policy proposal is a response to the recognition that consumer choices do contribute substantially to the cumulative level of GHGs discharged annually.

However, this proposal would impose high administrative burdens and private transaction costs to keep track of the climate change effects of many millions of people making billions of consumption choices every year.[434] I anticipate that this use of a disclosure mandate to track the consumption choices of Americans would lead to extreme political hostility and would succeed in uniting the Republican and Democratic Party leadership for the first time in decades. In essence, this proposed disclosure initiative is an attempt to assign a large share of the responsibility for overcoming climate change to decentralized consumers, which in my opinion would be an unrealistic and ineffectual approach that would almost certainly prove counter-productive from a political perspective.

Another type of mandatory disclosure requirement would make all major and mid-sized GHG pollution sources collect and publicize information about their current discharge levels and emissions-reduction efforts. This form of periodic information collection and disclosure would facilitate regulatory enforcement actions by the responsible agencies. Rather than making the agencies acquire most of the essential information themselves through independent monitoring efforts, US environmental regulatory statutes have almost always required substantial self-monitoring, data collection, and public disclosure by the regulated polluters; and many environmental laws have also imposed major penalties in the event of invalid disclosures or deliberate non-disclosures.

As one illustration of this disclosure mechanism applicable to regulated pollution sources, during 2009 and 2010 the EPA developed an evolving set of disclosure rules to require "reporting of greenhouse gas (GHG) emissions from large sources and suppliers in the United States. [40 CFR] Part 98 is intended to collect accurate and timely emissions data to inform future policy decisions."[435] The agency contends that: "EPA's GHG reporting system will provide a better understanding of where GHGs are coming from and will guide development of sound policies and programs to reduce emissions. This comprehensive, nationwide emissions data will help in the fight against climate change."[436]

Many other disclosure mechanisms are possible, and the specific requirements will have to be selected and refined by the responsible regulatory agencies. In early 2010, for example, the US Securities and Exchange Commission (SEC) acting on behalf of potential investors decided "to require publicly traded companies to disclose information regarding business risks and opportunities related to climate change."[437] These kinds of GHG disclosure programs should be regarded as overlapping mechanisms that can support the other mitigation institutions by creating stronger incentives for large GHG sources to reduce their discharges or to shift to GHG-free technologies in the near future.[438] These programs may also help all polluters, both consumers and businesses, make better choices about their own mitigation options. This discussion emphasizes that disclosure programs can help improve the performance of other kinds of mitigation programs, and in some GHG contexts they could serve as the primary mechanisms for promoting better climate-related decisions.

Climate change risks will affect all facets of human life and will require many complex and controversial climate-policy choices. Under these conditions, there is no reason to expect that only one type of regulatory institution could successfully undertake all necessary mitigation and adaptation measures.[439] To the contrary, several overlapping GHG regulatory regimes and economic incentive programs are likely to be more effective than any single approach could be.

CHAPTER VI

Conclusion

Will our descendants 100 or 200 years from now praise us for adopting thoughtful policies to limit climate change dangers, or will they sharply criticize us for failing to reduce the climate-related risks that our generation has been creating? Fierce condemnation appears far more likely than praise because the world's major GHG-polluting nations are continuing to increase their harmful discharges while international negotiations, national emissions-reduction commitments, and voluntary lifestyle changes have not begun to reverse this self-destructive behavior in any significant way. We cannot make real progress on climate change problems until our political leaders and climate experts recognize that the consensus emissions-reduction mitigation approaches are bound to fail while they continue allowing persistent residual GHG discharges to increase the cumulative atmospheric GHG concentration.

Global climate change has already arrived and is now creating numerous disasters for vulnerable people and ecosystems.[440] The heat-trapping climate contamination that is causing the atmospheric greenhouse effect and global warming has been documented by thousands of research findings with conclusions that are as close to scientific certainty as reputable scientists can come.[441] Unfortunately, nearly everything that concerned political leaders have promised to do to confront climate change dangers is being undermined by uncompromising ideological conflicts and the climate-policy mistakes described in this book. The inability or unwillingness of policymakers to understand the causes and consequences of climate change impacts is exposing billions of people to unprecedented social and ecological risks — exposures occurring every day and growing worse every year.

It is frustrating and nearly incomprehensible that the Republicans in the US Congress are almost unanimously opposed to adopting strong climate change regulations, and they are supported by a number of Democrats who represent fossil fuel-producing states. The US has suffered a wide range of climate-related disasters in the past two years, and many of the worst impacts occurred in states that continue to elect intransigent Republican legislators. Record or near-record floods have devastated many communities in the Mississippi River Valley and Missouri River Valley,[442] as well as most Southern states, including Alabama, Arkansas, Georgia, and Tennessee.[443] Unusually high temperatures are exacerbating heatwaves and wildfires causing terrible damage in Arizona, New Mexico, and California; and severe heatwaves in the spring and summer of 2011 also afflicted most of the Midwestern and Eastern states.[444] Powerful tornados caused hundreds of deaths and huge property losses in Southern and Midwestern states, and their lethal range extended as far east as Massachusetts.[445] Record snowfalls and snowpacks in the Western and Northwestern states are threatening massive flooding in those regions.[446] In contrast, vital water supplies have been severely limited by the worst drought in Texas in 50 years and rapidly decreasing water availability in other Southwestern and Southern states.[447] These damages are likely to be amplified because, as of summer 2011, many meteorologists and climatologists are predicting that there will be an unusually dangerous hurricane season this year.[448]

This is only a cursory summary of the recent human harms and economic damages in the US from extreme weather events. The unprecedented risks from climate change now, not only in the future, would appear even more threatening if descriptions were added here summarizing the recent disasters in vulnerable areas of developing countries and also the rapid changes in the Polar regions. Yet, many Congresspeople who should accept the responsibility for sponsoring effective climate mitigation programs are instead mired in unyielding "denial" despite the deaths and damages affecting their constituents. Although it is very difficult for scientists to attribute any particular extreme weather event to climate change, the prevalence of so many disastrous weather conditions in the same time periods and locations is exactly what thousands of climate scientists have been predicting for years.[449]

It is sad and shameful that the intransigence of many US politicians has been preventing the adoption of meaningful international climate change agreements and undermining the chance of ambitious mitigation initiatives by other developed nations. There is virtually no possibility that major GHG-polluting nations will accept strict, expensive precautionary programs needed to overcome climate change hazards without the US taking the lead. I am sure many readers want to know *why* — why is the US, the world's richest nation and the country most responsible for global warming over the past century, unwilling to accept a prominent role in finding ways to overcome the climate change dangers it has partly created? There is no single or simple answer to this fundamental problem, and any attempt to provide a serious answer is bound to entail considerable speculation. Nevertheless, here are several relevant factors that may contribute to a plausible answer:

- Republican members of Congress and their financial supporters have close connections to the American business community. Most business organizations complain about the high costs of climate change regulation and the job losses they claim will result, but these asserted losses pale into insignificance when compared against the widespread damages from extreme climate-related weather conditions the US and the world are already experiencing.[450]
- The fossil fuel industries that oppose effective GHG regulation (and economic incentives programs) are among the wealthiest financial forces on Earth, and they are not reluctant to deploy battalions of lobbyists, lawyers, bankers, and publicists wielding large political campaign contributions and other forms of bribery and corruption.[451]
- The American political system seems more and more dominated by short-term factors, while largely ignoring long-term consequences

that may prove harmful for US citizens. This short-sighted propensity applies not only to climate change risks, but to the recurring policy disputes on health care insurance, elder care, public education, immigration, infrastructure maintenance, and international security problems.

- Americans have one of the lowest national tax rates among developed nations, but they have been deluded into believing that they are horribly oppressed by all kinds of taxes and cannot afford more public expenditures for any purpose, no matter how vital. The opponents of strong climate mitigation efforts complain that these measures will require increased taxes and will impose higher costs on the American people, without acknowledging that the costs of *not overcoming* climate change dangers will be incomparably greater for a much longer period.[452]

- The US political system has become horribly polarized in recent decades. Whether conflicting views are expressed in terms of red states versus blue states, conservatives versus liberals, wealthy people versus workers, or Republicans versus Democrats, this opposition for the sake of opposition means that any climate-policy position adopted by "one side" will automatically be rejected by the "other side." As a result, Republican Congresspeople are close to unanimously opposed to climate change mitigation programs because the Democrats are almost unanimously in favor of climate change regulation unless they represent fossil fuel-producing states.

- Public-interest advocates cannot *fool* the people who hold strong ideological convictions by exposing them to scientifically proven facts. An interesting literature is emerging that shows ideology is a more powerful determinant of many people's beliefs than are objective facts or logic.[453] One context in which the primacy of ideology over knowledge and rational thought is important involves true believers in capitalism who are wholly devoted to "free markets" despite the fact that unrestricted market choices are precisely what has led to the rapid growth of greenhouse gas emissions and climate change harms.[454] In the view of these people, US society would be better off suffering through severe climate change disasters — which they doubt will actually occur despite the scientific consensus — than by adopting "socialist" climate regulation imposed by untrustworthy governments.

- A related psychological and social science literature has explored the prevalence of *denial* on "multiple levels, from emotions to cultural norms to political economy."[455] Most Americans do not want to accept their role as climate change villains, preferring more wasteful consumption to the alternative of shifting to sustainable living patterns. With two active wars, a record budget deficit, a frail economy still suffering from recession conditions, a high unemployment rate, and a widespread loss of personal and national security, it is hardly surprising that many Americans do not want to acknowledge the importance and urgency of climate mitigation needs.[456] In order to respond effectively and fairly to the need for reducing climate change risks, Americans would probably have to change their behaviors more than the people in any other country; and Americans would probably have to offer more financial assistance than any other donor nation to the climate change victims in less affluent countries who have been threatened by past and present US actions. Under these unpleasant circumstances, it cannot be surprising that a great many Americans and their political representatives are unwilling to face the human and environmental consequences of their actions or to recognize their responsibility for creating diverse climate change dangers.
- As many books supporting climate change mitigation have unintentionally revealed, *denial* is a two-edged sword. Environmentalists often will not acknowledge that the political opposition consists of more than only a few hundred greedy companies together with some corrupt politicians and the lobbyists that serve them. Dr. Hansen, for example, has blamed the lack of progress in promoting stronger climate policies on the influence of powerful private-interest groups and the lack of government vision,[457] which are certainly valid concerns, but he apparently has not accepted the

absence of broad popular support in America for self-sacrificing mitigation measures and the relatively low priority that many millions of US citizens have given to climate change problems.[458]

- After more than 60 years as the most successful and privileged people in the world, many Americans may have lost a sense of the need for major sacrifices to help each other and less fortunate people in many other countries. Climate change mitigation is generally described to the American people in terms of *what they must give up* in order to help save countless people and environmental features in other areas of the world. Yet, most Americans believe that they are entitled to retain a favorable position among nations as a result of their hard work and divine inspiration. Many people in the US are quite generous, probably more generous than the citizens of most other nations, but few Americans will choose to jeopardize their ultimate economic and social welfare by sacrificing their wealth and endowments for the benefit of others. As discussed in Chapter IV, this kind of radical sacrifice is exactly what many leaders in developing countries are demanding that the affluent developed nations should agree to make, and the predictable result has been an ongoing stalemate.

I see no realistic way to prevent the atmospheric GHG concentration and the greenhouse effect in the air from becoming steadily more harmful except by focusing on the development and dissemination of GHG-free replacement technologies and other "clean" methods that can promote greater economic and social development without causing progressively worse climate degradation. If we do not implement the necessary conversions to a clean GHG-free society now and in the near future, the persistent character of carbon discharges will ensure that billions of people must suffer greatly over the succeeding centuries. "*Too late*" is a terrible climate-policy prospect that many policymakers and expert advisors are still not fully appreciating.

The major GHG-polluting states still continue to discharge large quantities of persistent residual pollution that will increase the cumulative atmospheric GHG concentration until it reaches a level at which climate catastrophes are inescapable. As a result of reaching possibly irreversible "tipping points," we cannot be certain when disastrous climate changes will occur or how harmful they will be. Yet, all present and proposed consensus emissions-reduction programs that would allow large residual GHG emissions for decades are bound to be mitigation failures, not climate change solutions. This conclusion applies regardless of whether the back-loaded emissions-reduction programs employ direct regulation, cap-and-trade systems, carbon taxes, or other economic-incentives approaches. As long as emissions-reduction programs authorize annual discharges of persistent residual GHGs exceeding the amount of GHGs that natural or human-made sinks can remove from the atmosphere, these mitigation efforts cannot possibly overcome expanding climate change perils.

One of my academic friends gave a previous draft of this book to an Advisor in the UN Secretary-General's climate change office, and the Advisor reported that he liked the book very much but he wondered what my political solution would be for the national and international problems I have described. I do not have a realistic, comprehensive political solution today and neither does anyone else.

We need a profound political and economic transformation that will inspire a sincere global commitment to overcoming dangerous climate change risks. Most likely, this transformation will only come about as a result of truly catastrophic worldwide climate degradation, which will create massive human and ecological damages that are widely shared. This is obviously an unpleasant prospect, but rational discussions do not seem to be getting us anywhere at this time. There is no assurance that meaningful cooperation will replace the self-centered, self-defeating wrangling that now dominates both national and international climate-related proceedings. Despite the enormous amount of talking, writing, and negotiating on various climate issues, very little actual progress has occurred in the past two decades and the most severe climate problems continue growing worse, not better.

This book was not written to provide a comprehensive *political* solution, which does not exist now and is not on the horizon. Rather, the book was meant to focus on four central issues: First, it describes the conventional emissions-reduction programs, whether direct regulation or economic incentive mechanisms, and explains why these back-loaded mitigation efforts have uniformly failed in practice and will continue to fail. Second, the book presents an alternative approach focused on creating clean GHG-free replacement technologies and methods that have a more realistic chance to succeed technically and politically than the familiar emissions-reduction efforts. Third, the book explains the conflicting climate policies of Southern developing nations and affluent Northern states; and it emphasizes that the consensus emissions-reduction programs have offered virtually nothing that the Southern states want. In contrast, the clean replacement technology approach implemented through technology transfers and foreign aid contributions would allow the Southern countries to increase their economic development without further degrading the climate. Fourth and finally, the book contends that no single institutional solution can work sufficiently well by itself, including economic incentive systems, and that multiple institutional regimes have a significantly greater chance to succeed by incorporating overlapping mitigation measures that could help remedy the weaknesses of each separate institution.

Anyone serious about promoting climate change mitigation must understand many controversial issues and how to resolve them as effectively as possible. It may not be realistic now to attempt to convert this knowledge into immediate precautionary actions under current political conditions. And yet, if the world's leaders and climate experts are ever able to overcome the political paralysis that has been defeating climate-control efforts, these policymakers will need to understand which mitigation approaches can perform reasonably successfully and which approaches are certain to fail. That is the major contribution of this book.

Political conflicts and political power are transitory phenomena that can change rapidly in response to shifting external conditions and internal priorities. We can hope that many people in the near future will be more concerned with climate change dangers, and that more politicians will modify their positions on climate change accordingly. Global climate change might be largely ignored in the

2012 US Presidential election or it could become one of the most pressing issues in the campaign, depending on the environmental disasters and economic losses that occur during the next two years. We cannot predict when the majority of Americans will become seriously alarmed by imminent climate change risks, but the US political system and market system will most likely react to revised popular attitudes when they do emerge.

Proponents of climate mitigation must be prepared to "seize the moment" when evolving political currents begin to focus on the need for better climate change protections. We have to know what to do, and what not to do. This book is a roadmap pointing to thoughtful climate mitigation measures worth pursuing when the right time has arrived from a political perspective, and to foolish mitigation efforts that are wasting billions of dollars with no possibility of success.

Endnotes

1. Climate Change Poses "Defining Challenge" of Our Time, Ban Says (7 October 2008). UN News Service.
2. Obama, Barack (7 October 2008). Campaign Speech, Portsmouth, NH. http://www.barackobama.com/pdf/issues/EnvironmentFactSheet.pdf/ [10 November 2008].
3. Revkin, Andrew C. (16 September 2009). August Seas Warmest in at Least 120 Years. Dot Earth Blog. *New York Times*, Online Edition.
4. Broder, John M. (22 January 2010). Past Decade Warmest Ever, NASA Data Shows. *New York Times*, Online Edition.
5. *Id.*
6. Gillis, Justin (12 January 2011). Figures on Global Climate Show 2010 Tied 2005 as the Hottest Year on Record. *New York Times*, Online Edition.
7. Metz, Bert (2010). *Controlling Climate Change*. Cambridge: Cambridge University Press; Leonhardt, David (20 July 2010). Overcome by Heat and Inertia. *New York Times*, Online Edition; Rummukainen, Markku *et al.* (May 2010). Physical Climate Science Since IPCC AR4: A Brief Update on New Findings Between 2007 and April 2010. Nordic Council of Ministers. Norden website, http://www.norden.org/en/publications/publications/2010-549/ [25 June 2010]; Bhanoo, Sindya N. (29 January 2010). Less Water Vapor May Slow Warming Trends, *New York Times*, citing Susan Solomon *et al.* (28 January 2010). Contributions of Stratospheric Water Vapor to Decadal Changes in the Rate of Global Warming. *Science*. http://scienceonline.org/DOI:10.1126/science.1182488/ [29 January 2010]; Broder, John M. (22 January 2010). Past Decade Warmest Ever, NASA Data Shows. *New York Times*, Online Edition; Willett, M., L.V. Alexander and P.W.

Thorne (eds.) (June 2010). Global Climate. In *State of the Climate in 2009. Bulletin of the American Meteorological Society*, 91(6), S19–S28; Mann, Michael E. and Lee R. Kump (2008). *Dire Predictions: Understanding Global Warming*. London: DK Publishing.

With regard to media sources that claim many scientists dispute the existence or human causes of climate change, see the refutations in Hoggan, James with Richard Littlemore (2009). *Climate Cover-Up: The Crusade to Deny Global Warming.* Vancouver: Greystone Books; US EPA (3 August 2010). Denial of Petitions for Reconsideration of the Endangerment and Cause or Contribute Findings for Greenhouse Gases Under Section 202(a) of the Clean Air Act; Associated Press (30 March 2010). UK Panel Call Climate Data Valid. *New York Times,* Online Edition; Pew Center on Global Climate Change (29 October 2009). Realities vs. Misconceptions About the Science of Climate Change.

8. Metz, Bert (2010). *Controlling Climate Change,* pp. 12–29. Cambridge: Cambridge University Press; Cullen, Heidi (2010). *The Weather of the Future: Heat Waves, Extreme Storms and Other Scenes from a Climate-Changed Planet*. New York: Harper Collins; Flannery, Tim (2009). *Now or Never: Why We Must Act Now to End Climate Change and Create a Sustainable Future*. New York: Atlantic Monthly Press; Kolbert, Elizabeth (2006). *Field Notes from a Catastrophe: Man, Nature, and Climate Change.* New York: Bloomsbury; Backlund, P., A. Janetos, D. Schimel *et al.* (May 2008). The Effects of Climate Change on Agriculture, Land Resources, Water Resources, and Biodiversity in the United States: Synthesis and Assessment Product 4.3. US Climate Change Science Program Subcommittee on Global Change Research; Intergovernmental Panel on Climate Change (December 2007). Fourth Assessment Report on Climate Change 2007: Synthesis Report — Summary for Policymakers; Nepstad, Daniel C. *et al.* (11 February 2008). Interactions Among Amazon Land Use, Forests and Climate: Prospects for a Near-Term Forest Tipping Point. *Philosophical Transactions of the Royal Society B*, 363, 1737–1746; Dasgupta, Susmita *et al.* (February 2007). The Impact of Sea Level Rise on Developing Countries: A Comparative Analysis. World Bank Policy Research Working Paper 4136; van Aalst, Maarten (2006). Managing Climate Risk: Integrating Adaptation into World Bank Group Operations. World

Bank Group, Global Environmental Facility Program; Schipper, Lisa and Mark Pelling (2006). Disaster Risk, Climate Change and International Development: Scope for, and Challenges to, Integration. *Disasters*, 30(1), 19–38; Gitay, H., A. Suárez, R. Watson and D.J. Dokken (eds.) (April 2002). Climate Change and Biodiversity. Intergovernmental Panel on Climate Change. IPCC Technical Paper V.

9. Hansen, James (2009). *Storms of My Grandchildren: The Truth About the Coming Climate Catastrophe and Our Last Chance to Save Humanity*. New York: Bloomsbury [hereafter cited as *Storms of My Grandchildren*]; Orr, David W. (2009). *Down to the Wire: Confronting Climate Collapse*. Oxford: Oxford University Press; Brown, Lester R. (2008). *Plan B: 3.0*. London: W.W. Norton & Company; Friedman, Thomas L. (2008). *Hot, Flat, and Crowded: Why We Need a Green Revolution — And How It Can Renew America*. New York: Farrar, Straus and Giroux; Monbiot, George (2007). *Heat: How to Stop the Planet from Burning*. Cambridge, MA: South End Press; Pearce, Fred (2008). *With Speed and Violence*. Boston: Beacon Press; Ward, Peter D. (2007). *Under a Green Sky*. New York: Harper Collins; Hansen, Jim (13 July 2006). The Threat to the Planet. *New York Review of Books*, 53(12), 13; Sorkin, Lauren (2006–2007). Climate Change Impacts Rise. *Vital Signs*, pp. 42–43, 133–134. Worldwatch Institute.
10. Revkin, Andrew C. (18 November 2008). Obama: Climate Plan Firm Amid Economic Woes. Dot Earth Blog. *New York Times*, Online Edition.
11. Kommareddi, Madhuri (2008). Barack Obama for President Campaign: Environment Fact Sheet. http://www.barackobama.com/pdf/issues/EnvironmentFactSheet.pdf/ [29 December 2008].
12. UN Framework Convention on Climate Change (1992). *International Legal Materials*, 31, 849. Concluded at Rio de Janeiro on 29 May 1992. Entered into force on 21 March 1994.
13. Fahrenthold, David A. (18 December 2009). 8 Questions About the Last Day of Copenhagen Climate Talks. *Washington Post*, Online Edition.
14. Stern, Todd (28 January 2010). Letter to Yvo de Boer, Executive Secretary, United Nations Framework Convention on Climate Change.
15. Hulse, Carl and David M. Herszenhorn (22 July 2010). Democrats Call Off Climate Bill Effort. *New York Times*, Online Edition.
16. US House of Representatives (March 2009). Discussion Draft Summary of the American Clean Energy and Security Act of 2009;

Broder, John M. (1 April 2009). Democrats Unveil Climate Bill. *New York Times*, Online Edition.

17. Kemp, John (27 May 2009). US Climate Change Bill: Radically Bad Law. Reuters website, http://www.reuters.com/article/idUSTRE54Q48C20090527/ [1 June 2009]; Galbraith, Kate (22 May 2009). The Climate Debate Continues. *New York Times*, Online Edition; Handouts and Loopholes: America's Climate-Change Bill Is Weaker and Worse than Expected (21 May 2009). *The Economist*; Krugman, Paul (18 May 2009). The Perfect, the Good, the Planet. *New York Times*, Online Edition.
18. Pooley, Eric (2010). *The Climate War: True Believers, Power Brokers and the Fight to Save the Earth*. New York: Hyperion.
19. Tankersley, Jim (6 July 2009). Climate Battle Moves to the Senate: Obama Faces Calls for More Concessions, Including Offshore Drilling, in Global Warming Legislation. *Los Angeles Times*, Online Edition; Broder, John M. (27 June 2009). House Passes Bill to Address Threat of Climate Change. *New York Times*, Online Edition.
20. Broder, John M. (17 May 2009). From a Theory to a Consensus on Emissions. *New York Times*, Online Edition.
21. 111th Congress, 2d Session (2010). American Power Act of 2010. Discussion Draft; Broder, John M. (12 May 2010). Senate Gets a Climate and Energy Bill, Modified by a Gulf Spill That Still Grows. *New York Times*, Online Edition.
22. Broder, John M. (7 May 2010). Graham Calls for 'Pause' in Pursuing Energy Bill. *New York Times*, Online Edition; Eilperin, Juliet (25 April 2010). Graham Withdraws Support for Climate Legislation. *Washington Post*, p. A03.
23. US Climate Bill Emerges for Uphill Battle (13 May 2010). Carbon Positive website, http://www.carbonpositive.net/viewarticle.aspx?articleID=1998/ [6 November 2010]; The Climate Change Bill: Once More unto the Breach (13 May 2010). *The Economist*; Room for Debate: Does the Climate Bill Have a Chance? (9 May 2010). *New York Times*, Online Edition; Kitasei, Saya (26 March 2010). What Is the Real Sense of the Senate? ReVolt: The Worldwatch Institute's Climate and Energy Blog website, http://blogs.worldwatch.org/revolt/what-is-the-real-sense-of-the-senate/ [30 April 2010].
24. 111th Congress, 2d Session (2010). American Power Act of 2010, Sec. 702. Discussion Draft; Lashof, Dan (13 May 2010). Solid at the Core:

The Integrity of the Emission Limits in the American Power Act. Switchboard: NRDC Staff Blog website, http://switchboard.nrdc.org/blogs/dlashof/ [15 May 2010].

25. Stavins, Robert (18 May 2010). Here We Go Again: A Closer Look at the Kerry–Lieberman Cap-and-Trade Proposal. Harvard University Belfer Center for Science and International Affairs website, http://belfercenter.ksg.harvard.edu/analysis/stavins/?p=643/; Schmidt, Jake (13 May 2010). Tools for Supporting International Action on Global Warming: American Power Act. Switchboard: NRDC Staff Blog website, http://switchboard.nrdc.org/blogs/jschmidt/apa_intl_provisions.html/ [15 May 2010]; US Climate Bill to Set Utility Cap-Trade (15 March 2010). *Washington Post*, Online Edition.
26. California Air Resources Board (October 2008). Climate Change: Proposed Scoping Plan; Broder, John M. and Peter Baker (26 January 2009). Obama's Order Is Likely to Tighten Auto Standards. *New York Times*, Online Edition.
27. Broder, John M. (15 September 2009). New Standard Links Auto Mileage and Gas Emissions. *New York Times*, Online Edition.
28. Vlasic, Bill (28 July 2011). Carmakers Back Strict New Rules for Gas Mileage. *New York Times*, Online Edition.
29. Regional Greenhouse Gas Initiative website, http://www.rggi.org/ [20 August 2008].
30. The quoted passage comes from the following press release: US States, Canadian Provinces Announce Regional Cap-and-Trade Program to Reduce Greenhouse Gases (28 September 2008). WCI website, http://www.westernclimateinitiative.org/ewebeditpro/items/O104F19871.pdf/.
31. Kyoto Protocol to the United Nations Framework Convention on Climate Change (10 December 1997). Reprinted in *International Legal Materials*, 37, 22; Danish, Kyle W. (2007). An Overview of the International Regime Addressing Climate Change. *Sustainable Law & Policy*, 7, 10.
32. Okereke, Chukwumerije *et al.* (June 2007). Assessment of Key Negotiating Issues at Nairobi Climate COP/MOP and What It Means for the Future of the Climate Regime. Tyndall Centre for Climate Change Research Working Paper 106; Olmstead, Sheila M. and Robert N. Stavins (March 2006). An International Architecture for the Post-Kyoto Era. Harvard University John F. Kennedy School of Government Faculty Working Paper RWP06-009.

33. Winds of Change: The EU Unveils Bold Plans to Tackle Global Warming (9 March 2007). *The Economist*, Online Edition; Herro, Alana (17 July 2006). Kyoto: Impossible Goal or Economic Opportunity? Worldwatch Institute website, http://www.worldwatch.org/node/4362/ [27 September 2006].
34. Cutting Carbon in Europe: The 2020 Plans and the Future of the ETS (6 January 2008). Carbon Trust website, http://www.carbontrust.co.uk/Publications/pages/publicationdetail.aspx?id=ctc734/ [1 January 2009]; EU Reaches Final 2020 Climate Deal (15 December 2008). Carbon Positive website, http://www.carbonpositive.net/viewarticle.aspx?articleID=1340/; The EU Summit: Keeping It Clean (12 December 2008). *The Economist*, Online Edition.
35. Mickwitz, Per *et al.* (2009). Climate Policy Integration, Coherence and Governance. Partnership for European Environmental Research Report No. 2. PEER website, http://www.peer.eu/fileadmin/user_upload/publications/PEER_Report2.pdf/ [3 April 2009]; Carbon Trust (2008). Cutting Carbon in Europe: The 2020 Plans and the Future of the EU ETS. Carbon Trust website, http://www.carbontrust.co.uk/Publications/pages/publicationdetail.aspx?id=ctc734/ [1 January 2009]; Kanter, James and Stephen Castle (12 December 2008). European Leaders Agree on Weakened Plan to Reduce Emissions. *New York Times*, Online Edition.
36. Climate Change: Commission Invites to an Informed Debate on the Impacts of the Move to 30% EU Greenhouse Gas Emissions Cut If and When the Conditions Are Met (26 May 2010). EUROPA Press Release. European Union Portal IP/10/618; European Commission (2010). Communication from the Commission to the Council, the European Parliament, the European Economic and Social Committee and the Committee of the Regions (Unofficial Version): Analysis of Options to Move Beyond 20% Greenhouse Gas Emission Reductions and Assessing the Risk of Carbon Leakage. COM 265/3.
37. Rosenthal, Elisabeth (1 July 2010). Britain Curbing Airport Growth to Aid Climate. *New York Times*, Online Edition.
38. Beament, Emily (7 October 2008). UK Must Cut Greenhouse Gases by 80 Percent, Committee Says. *The Independent* website, http://www.independent.co.uk/environment/climate-change/uk-should-cut-greenhouse-gases-by-80-per-cent-953797.html/.

39. *Id.*
40. Harvey, Fiona and Allegra Stratton (17 May 2011). Chris Huhne Pledges to Halve UK Carbon Emissions by 2025. *The Guardian*, Online Edition.
41. Natural Resources Defense Council (April 2008). Solicitation Letter to the author in April 2008.
42. Environmental Defense Fund (29 February 2008). Solicitation Letter, 2008 Climate Action Plan.
43. Union of Concerned Scientists (26 February 2008). Global Warming 101: A Target for US Emissions Reductions. Email to author.
44. Solomon, Susan *et al.* (10 February 2010). Irreversible Climate Change Due to Carbon Dioxide Emissions. *Proceedings of the National Academy of Sciences*, 106(6), 1704–1709; Revkin, Andrew C. (28 January 2009). The Greenhouse Effect and the Bathtub Effect. Dot Earth Blog. *New York Times*, Online Edition; Dean, Cornelia (26 January 2009). Emissions Cut Won't Bring Quick Relief, Scientists Say. *New York Times*, Online Edition.
45. Metz, Bert (2010). *Controlling Climate Change*, pp. 8–9, 31–33, 42–49. Cambridge: Cambridge University Press; Emanuel, Kerry (2007). *What We Know About Climate Change*, pp. 15–21. Cambridge, MA: MIT Press; Rosenzweig, C. *et al.* (2007). Assessment of Observed Changes and Responses in Natural and Managed Systems. In M.L. Parry *et al.* (eds.). *Climate Change 2007: Impacts, Adaptation and Vulnerability. Contribution of Working Group II to the Fourth Assessment Report of the Intergovernmental Panel on Climate Change*, pp. 79–131. Cambridge: Cambridge University Press; Climate Change Science: State of Knowledge (2007). US Environmental Protection Agency website, http://www.epa.gov/climatechange/science/stateofknowledge.html/ [29 April 2007].
46. As used in this discussion, "clean" means any process that does not directly or indirectly generate greenhouse gases. I am not using "clean" to encompass all of the other pollutants and toxic substances produced by our industrial and consumption activities.
47. For a similar conclusion, see Sachs, Jeffrey D. (April 2008). Keys to Climate Protection. *Scientific American*, 298(4), 40; —— (March 2008). Climate Change After Bali. *Scientific American*, 298(3), 33–34.

48. Predicting a World Population of 9.3 Billion in 2050 (18 June 2008). US Census Bureau International Data Base; World Population to Reach 9.1 Billion in 2050, UN Projects (24 February 2005). UN News Centre.
49. US Environmental Protection Agency (2007). Future Climate Change. US EPA website, http://www.epa.gov/climatechange/science/futurecc.html/ [23 April 2007]; Intergovernmental Panel on Climate Change (February 2007). Stabilization of Atmospheric Greenhouse Gases: Physical, Biological and Socio-economic Implications; Timilsina, Govinda R. (September 2007). Atmospheric Stabilization of CO_2 Emissions: Near-Term Reductions and Intensity-Based Targets. World Bank Policy Research Working Paper 4352.
50. Pielke, Roger, Jr., Tom Wigley and Christopher Green (3 April 2008). Dangerous Assumptions. *Nature*, 452, 531–532; Kristoff, Nicholas D. (20 April 2008). Our Favorite Planet. *New York Times*, Online Edition.
51. Ban, Ki Moon (13 June 2008). Securing the Common Good: The United Nations and the Expanding Global Agenda. UNA-UK address, London. UN News Service.
52. Metz, Bert (2010). *Controlling Climate Change*, pp. 1–9. Cambridge: Cambridge University Press; Intergovernmental Panel on Climate Change (2007). Climate Change: The Physical Science Basis. Contribution of Working Group I to the Fourth Assessment Report, pp. 129–234, 755–759; —— (2007). Summary for Policymakers. In S. Solomon *et al.* (eds.). *Climate Change 2007: The Physical Science Basis. Contribution of Working Group I to the Fourth Assessment Report of the Intergovernmental Panel on Climate Change*. Cambridge: Cambridge University Press; Emanuel, Kerry (2007). *What We Know About Climate Change*, pp. 15–21. Cambridge, MA: MIT Press.
53. Canadell, J. *et al.* (17 September 2007). Contributions to Accelerating Atmospheric CO_2 Growth from Economic Activity: Carbon Intensity, and Efficiency of Natural Sinks. *Proceedings of the National Academy of Sciences* website, http://www.pnas.org/content/104/47/18866.full.pdf/ [14 November 2007]; Climate Change Science: State of Knowledge (2007). US Environmental Protection Agency website, http://www.epa.gov/climatechange/science/stateofknowledge.html/ [29 April 2007]; Environment Canada (April 2001). The Science of Climate Change — Part I: Introduction/GHG/Radiative Forcing; Revkin, Andrew C.

(4 May 2007). Climate Panel Reaches Consensus on the Need to Reduce Harmful Emissions. *New York Times*, Online Edition.

54. Pacala, S. and R. Socolow (13 August 2004). Stabilization Wedges: Solving the Climate Problem for the Next 50 Years with Current Technologies. *Science*, 305(5686), 968–972.
55. Regional Greenhouse Gas Initiative website, http://www.rggi.org/ [20 August 2008].
56. Navarro, Mireya (26 May 2011). Christie Pulls New Jersey from 10-State Climate Initiative. *New York Times*, Online Edition.
57. *Id.*
58. *Id.*
59. *Id.*
60. Sperling, Daniel and Deborah Gordon (2009). *Two Billion Cars: Driving Toward Sustainability*. New York: Oxford University Press.
61. Revkin, Andrew C. (10 July 2008). After Applause Dies Down, Global Warming Talks Leave Few Concrete Goals. *New York Times*, Online Edition; Stolberg, Sheryl Gay (9 July 2008). G-8 Leaders Pledge to Cut Emissions in Half by 2050. *New York Times*, Online Edition.
62. Dr. Sterman's comments are quoted in Revkin, Andrew C. (28 January 2009). The Greenhouse Effect and the Bathtub Effect. *New York Times*, Online Edition.
63. Hodas, David (2008). Imagining the Unimaginable: Reducing U.S. Greenhouse Gas Emissions by Forty Percent. *Virginia Environmental Law Journal*, 26, 271.
64. The mileage and GHG figures were copied from Fueleconomy.gov website, http://www.fueleconomy.gov/feg/findacar.htm/ [13 September 2010].
65. Sackmann, Hendrik (14 September 2010). Carmakers Look to Deepen Cooperation on Electric Cars. *New York Times*, Online Edition; Bunkley, Nick (13 September 2010). Toyota Plans Six New Hybrids for 2012. *New York Times*, Online Edition; Martin, Christopher (3 August 2010). Electric Vehicles Capture Clean Energy Venture Capital, Ernst & Young Says. Bloomberg website, http://www.bloomberg.com/news/2010-08-03/electric-vehicles-capture-clean-energy-venture-capital-ernst-young-says.html/ [3 August 2010].
66. This transformation to all-electric cars would impose a greater demand for electricity generation that must be taken into account in

considering the cumulative GHG emissions-reduction results. However, this trade-off will have to be done for nearly every substantial reduction in the use of fossil fuels, and can be at least partly offset by better vehicle designs including solar cells and brake rotation to help recharge the vehicle batteries. For a general discussion of the many design options that can improve motor vehicles from a climate change perspective, see Sperling, Daniel and Deborah Gordon (2009). *Two Billion Cars: Driving Toward Sustainability.* Oxford: Oxford University Press.

67. Behr, Peter (9 September 2010). Forthcoming Study Finds Gap Between Expectations and Reality of Electric Cars. *New York Times*, Online Edition.
68. Climate Change Science: State of Knowledge (2007). US Environmental Protection Agency website, http://www.epa.gov/climatechange/science/stateofknowledge.html/ [29 April 2007].
69. Intergovernmental Panel on Climate Change (October 1997). Implications of Proposed CO_2 Emissions Limitations, p. 3.
70. House, Joanna I. *et al.* (2008). What Do Recent Advances in Quantifying Climate and Carbon Cycle Uncertainties Mean for Climate Policy? *Environmental Research Letters*, 3(044002), 1–6; Hansen, James (23 June 2008). Guest Opinion: Global Warming Twenty Years Later. *Worldwatch Magazine* website, http://www.worldwatch.org/node/5798/ [27 June 2008].
71. Ledley, Tamara S. *et al.* (28 September 1999). Climate Change and Greenhouse Gases. *Eos*, 80(39), 453.
72. Solomon, Susan *et al.* (10 February 2009). Irreversible Climate Change Due to Carbon Dioxide Emissions. *Proceedings of the National Academy of Sciences*, 106(6), 1704–1709; Inman, Mason (December 2008). Carbon Is Forever. *Nature Reports* website, http://www.nature.com/reports/climatechange/ [3 December 2008]; Eilperin, Juliet (10 March 2008). Carbon Output Must Near Zero to Avert Danger, New Studies Say. *Washington Post*, Online Edition; Matthews, H.D. and K. Caldeira (14 February 2008). Stabilizing Climate Requires Near-Zero Emissions. *Geophysical Research Letters*; Schmittner, Andreas *et al.* (14 February 2008). Future Changes in Climate, Ocean Circulation, Ecosystems, and Biogeochemical Cycling Simulated for a Business-as-Usual CO_2

Emission Scenario Until Year 4000 AD. *Global Biogeochemical Cycles*, 22, 1–21.

73. House, Joanna I. *et al*. (2008). What Do Recent Advances in Quantifying Climate and Carbon Cycle Uncertainties Mean for Climate Policy? *Environmental Research Letters*, 3(044002), 1.
74. Committee on Health, Environmental, and Other External Costs and Benefits of Energy Production and Consumption *et al*. (2010). *Hidden Costs of Energy: Unpriced Consequences of Energy Production and Use*. Washington, D.C.: National Academies Press.
75. Aldy, Joseph E. *et al*. (May 2009). Designing Climate Mitigation Policy. Resources for the Future Discussion Paper DP 08-16.
76. Metz, Bert (2010). *Controlling Climate Change*, p. 36. Cambridge: Cambridge University Press.
77. Morales, Alex (3 August 2010). Climate Loopholes May Wipe Out Greenhouse Gas Cuts, Island Nations Say. Bloomberg website, http://www.bloomberg.com/news/2010-08-03/climate-loopholes-may-wipe-out-greenhouse-gas-cuts-island-nations-say.html/ [3 August 2010].
78. European Commission (26 May 2010). Climate Change: Commission Invites to an Informed Debate on the Impacts of the Move to 30% EU Greenhouse Gas Emissions Cut If and When the Conditions Are Met. EUROPA website, http://europa.eu/rapid/pressReleasesAction.do?reference=IP/10/618&format=HTML&aged=0&language=EN&guiLanguage=en/.
79. Hansen, James (2009). *Storms of My Grandchildren*, p. 177.
80. *Id*.
81. Ramanathan, V. and Y. Feng (24 July 2008). On Avoiding Dangerous Anthropogenic Interference with the Climate System: Formidable Challenges Ahead. *Proceedings of the National Academy of Sciences* website, http://www.pnas.org/content/early/2008/09/16/0803838105.full.pdf/ [5 April 2010].
82. Understanding and Responding to Climate Change: Highlights of National Academies Reports (2008).
83. Pachauri, R.K. and A. Reisinger (eds.) (2008). Intergovernmental Panel on Climate Change Fourth Assessment Report. Climate Change 2007: Synthesis Report. Geneva; Canadell, J. *et al*. (17 September 2007).

Contributions to Accelerating Atmospheric CO_2 Growth from Economic Activity: Carbon Intensity, and Efficiency of Natural Sinks. *Proceedings of the National Academy of Sciences* website, http://www.pnas.org/content/104/47/18866.full/ [14 November 2007]; Associated Press (17 May 2007). Oceans May Be Losing Ability to Soak Up CO_2. *New York Times*, Online Edition.

84. Bhanoo, Sindya N. (19 November 2009). Seas Grow Less Effective at Absorbing Emissions. *New York Times*, Online Edition.
85. UN Department of Economic and Social Affairs (December 2009). Ocean Acidification: A Hidden Risk for Sustainable Development. Copenhagen Policy Brief No. 1.
86. Laffoley, Dan (26 December 2009). To Save the Planet, Save the Seas. *New York Times*, Online Edition.
87. Persson, U. Martin and Christian Azar (2010). Preserving the World's Tropical Forests: A Price on Carbon May Not Do. *Environmental Science and Technology*, 44(1), 210–215; Secretariat of the Convention on Biological Diversity (26 October 2009). Forest Biodiversity Provides an "Insurance Policy" Against Climate Change. Convention on Biological Diversity Press Release; McAlpine, Jan L. (17 June 2009). Forests and Climate Change: Bridging the Gaps. United Nations Forum on Forests (UNFF) Secretariat.
88. Canadell, J. *et al.* (17 September 2007). Contributions to Accelerating Atmospheric CO_2 Growth from Economic Activity: Carbon Intensity, and Efficiency of Natural Sinks. *Proceedings of the National Academy of Sciences* website, http://www.pnas.org/content/104/47/18866.full.pdf/ [14 November 2007].
89. Galbraith, Kate (13 June 2009). US–Private Bid to Trap Carbon Emissions Is Revived. *New York Times*, Online Edition; US Department of Energy (2007). Carbon Sequestration Technology Roadmap and Program Plan; Fischedick, Manfred *et al.* (2007). CO_2-Capture and Geological Storage as a Climate Policy Option. Wuppertal Institute Special Report 35e; IPCC Working Group III (2005). Special Report on Carbon Capture and Storage: Summary for Policymakers.
90. McKinsey & Co. and The Conference Board (December 2007). Reducing US Greenhouse Gas Emissions: How Much at What Cost?; International Petroleum Industry Environmental Conservation

Association (IPIECA) (March 2007). Oil and Natural Gas Industry Guidelines for Greenhouse Gas Reduction Projects; Intergovernmental Panel on Climate Change (2005). Carbon Dioxide Capture and Storage.

91. Wald, Matthew L. and John M. Broder (13 July 2011). Utility Shelves Ambitious Plan to Limit Carbon. *New York Times*, Online Edition.
92. *Id.*
93. *Id.*
94. Stolberg, Sheryl Gay (17 April 2008). Bush Sets Greenhouse Gas Emissions Goal. *New York Times*, Online Edition.
95. Collins, Gail (19 April 2008). The Fat Bush Theory. *New York Times*, Online Edition.
96. *Id.*
97. Henson, Robert (January 2008). *The Rough Guide to Climate Change, 2nd Edition*. London: Rough Guides Ltd.
98. Revkin, Andrew C. (1 December 2008). CO_2's Long Goodbye. Dot Earth Blog. *New York Times*, Online Edition.
99. Revkin, Andrew C. (28 January 2009). The Greenhouse Effect and the Bathtub Effect. *New York Times*, Online Edition.
100. *Id.*
101. Kunzig, Robert (December 2009). The Big Idea: The Carbon Bathtub. *National Geographic*, 216(6), 26–28.
102. Sitch, Stephen *et al.* (16 August 2007). Indirect Radiative Forcing of Climate Change Through Ozone Effects on the Land-Carbon Sink. *Nature*, 448, 791–794.
103. Union of Concerned Scientists (26 February 2008). Global Warming 101: A Target for US Emissions Reductions. Email message to author.
104. Union of Concerned Scientists (11 December 2007). Scientists Weigh in at Climate Negotiations in Bali. Email message to author.
105. Ekwurzel, Brenda (8 May 2009). UCS email message to the author, emphasis added.
106. Joeri, Rogelj (July 2009). Halfway to Copenhagen, No Way to 2°C. *Nature Reports Climate Change*, Vol. 3. *Nature* website, http://www.nature.com/climate/2009/0907/full/climate.2009.57.html/ [16 October 2009]; Meinshausen, Malte *et al.* (30 April 2009). Greenhouse-Gas Emission Targets for Limiting Global Warming to 2°C. *Nature Letters*, 458, 1158.

107. Associated Press (8 July 2009). G-8 Agrees to Cap on Global Temperatures But Developing Nations Fail to Commit to Specific Greenhouse Gas Cuts. MSNBC website, http://www.msnbc.msn.com/id/31802183/print/1/displaymode/1098/ [8 July 2009].
108. *Id.*
109. Geden, Oliver (August 2010). What Comes After the Two-Degree Target? SWP Comments 19.
110. The complete text of the Copenhagen Accord can be downloaded from the UNFCCC website at the following URL: http://unfccc.int/files/meetings/cop_15/application/pdf/cop15_cph_auv.pdf/.
111. Schneider, S.H. *et al.* (2007). Assessing Key Vulnerabilities and the Risk from Climate Change. In M.L. Parry *et al.* (eds.). *Climate Change 2007: Impacts, Adaptation and Vulnerability*, p. 781. Cambridge: Cambridge University Press.
112. *Id.*, p. 792.
113. *Id.*, p. 796.
114. Yohe, G.W. *et al.* (2007). Perspectives on Climate Change and Sustainability. In M.L. Parry *et al.* (eds.). *Climate Change 2007: Impacts, Adaptation and Vulnerability*, p. 825. Cambridge: Cambridge University Press.
115. Revkin, Andrew C. (10 August 2010). Scientists See Links from Asian Floods to Russian Heat. *New York Times*, Online Edition.
116. New, Mark, Diana Liverman and Kevin Anderson (December 2009). Commentary: Mind the Gap. *Nature Reports Climate Change*, Vol. 3. *Nature* website, http://www.nature.com/climate/2009/0912/full/climate.2009.126.html/ [22 December 2009].
117. Intergovernmental Panel on Climate Change (2007). Fourth Assessment Report. Synthesis Report: Summary for Policymakers, p. 22.
118. van Vuuren, Detlef, A.F. Hof and M.G.J. den Elzen (December 2009). Meeting the 2°C Target: From Climate Objective to Emission Reduction Measures. Netherlands Environmental Assessment Agency (PBL) publication number 500114012, pp. 11–12.
119. *Id.*, p. 11 (recommending the overshoot increase be held to no greater than 500 ppm).
120. Hamilton, Clive (21 October 2009). Is It Too Late to Prevent Catastrophic Climate Change? Lecture at a Meeting of the Royal Society of the Arts, Sydney, Australia, pp. 17–18.

121. Joeri, Rogelj (29 September 2010). Analysis of the Copenhagen Accord Pledges and Its Global Climatic Impacts — A Snapshot of Dissonant Ambitions. *Environmental Research Letters*, 5(034013), 1–2; Schleich, Joachim, Vicki Duscha and Everett B. Peterson (July 2010). Environmental and Economic Effects of the Copenhagen Pledges and More Ambitious Emission Reduction Targets: Interim Report. German Federal Ministry of the Environment website, http://www.uba.de/uba-info-medien/3998.html/; Eilperin, Juliet and Anthony Faiola (19 December 2009). Climate Deal Falls Short of Key Goal. *Washington Post*, p. A01.
122. Gurría, Angel (22 April 2010). Climate Change and Competitiveness. Organisation for Economic Co-operation and Development.
123. Patel, Urjit R. (March 2010). Decarbonisation Strategies: How Much, How, Where and Who Pays for $\Delta \leq 2°C$? Brookings Global Economy & Development Working Paper 39, p. 3; van Vuuren, Detlef, A.F. Hof and M.G.J. den Elzen (December 2009). Meeting the 2°C Target: From Climate Objective to Emission Reduction Measures. Netherlands Environmental Assessment Agency (PBL) publication number 500114012.
124. Hamilton, Clive (21 October 2009). Is It Too Late to Prevent Catastrophic Climate Change? Lecture at a Meeting of the Royal Society of the Arts, Sydney, Australia.
125. Geden, Oliver (August 2010). What Comes After the Two-Degree Target? SWP Comments 19, pp. 2–3.
126. *Id.*, pp. 3–4.
127. *Id.*, p. 4.
128. Intergovernmental Panel on Climate Change (2007). Fourth Assessment Report. Synthesis Report: Summary for Policymakers, p. 22.
129. The cap-and-trade conception is generally ascribed to the Canadian economist, John H. Dales. See Dales, J.H. (1968). *Pollution, Property and Prices: An Essay in Policy-Making and Economics.* Toronto: University of Toronto Press.
130. US Environmental Protection Agency (2004). Cap-and-Trade Essentials. US EPA website, http://www.epa.gov/capandtrade/documents/ctessentials.pdf/ [6 July 2008]; Stavins, Robert (November 2007).

Addressing Climate Change with a Comprehensive US Cap-and-Trade System. Harvard University JFK School of Government Faculty Research Working Paper RWP07-053; Schultz, Abby (13 April 2008). Sizing Up the Utilities, If Carbon Caps Take Hold. *New York Times*, Online Edition; Baldwin, Robert (2008). Regulation Lite: The Rise of Emissions Trading. London School of Economics Working Paper 3/2008. Social Science Research Network website, http://papers.ssrn.com/sol3/papers.cfm?abstract_id=1091784/ [6 June 2008].

131. Stavins, Robert (3 June 2009). An Economic View of the Environment — The Wonderful Politics of Cap-and-Trade: A Closer Look at Waxman–Markey. Harvard University Belfer Center for Science & International Affairs Working Paper; Keohane, Nathaniel O. (2009). Cap and Trade, Rehabilitated: Using Tradable Permits to Control US Greenhouse Gases. *Review of Environmental Economics & Policy*, 3(1), 42–62; Parsons, John E., A. Denny Ellerman and Stephan Feilhauer (2009). Designing a US Market for CO_2. *Journal of Applied Corporate Finance*, 21(1), 79–86; Stavins, Robert N. (September 2008). Addressing Climate Change with a Comprehensive US Cap-and-Trade System. Fondazione Eni Enrico Mattei Note di Lavoro Series No. 67.2008.
132. Driesen, David M. (2010). Capping Carbon. *Environmental Law*, 40(1), 1–51.
133. Kommareddi, Madhuri (2008). Barack Obama for President Campaign: Environment Fact Sheet. http://www.barackobama.com/pdf/issues/EnvironmentFactSheet.pdf/ [29 December 2008].
134. Hansen Appeals to Obama on Climate (7 January 2009). Carbon Positive website, http://www.carbonpositive.net/viewarticle.aspx?articleID=1358/.
135. Revkin, Andrew C. (14 October 2008). George Soros: New Energy = New Prosperity. *New York Times*, Online Edition.
136. Sokol, David (19 May 2009). Let's Have Cap and No Trade. *Washington Post*, Online Edition.
137. Friedman, Thomas L. (7 April 2009). Show Us the Ball. *New York Times*, Online Edition.
138. Morford, Stacy (21 May 2009). Climate Bill Wins Enough Votes to Pass, But at What Cost? Solve Climate website, http://solveclimate.

com/blog/20090521/climate-bill-wins-enough-votes-pass-what-cost/ [1 June 2009].

139. *Id.*; Kemp, John (27 May 2009). US Climate Change Bill: Radically Bad Law. Reuters website.
140. The Senate Proposals (28 March 2010). *Washington Post*, Online Edition.
141. Broder, John M. (25 March 2010). 'Cap and Trade' Loses Its Standing as Energy Policy of Choice. *New York Times*, Online Edition.
142. *Id.*
143. Feldstein, Martin (1 June 2009). Cap-and-Trade: All Cost, No Benefit. *Washington Post*, Online Edition.
144. *Id.*
145. Environmental Defense Fund (7 August 2008). Cap-and-Trade 101. Environmental Defense Fund website [1 January 2009].
146. 111th Congress, 2d Session (2010). American Power Act of 2010, §721(e). Discussion Draft.
147. Alberola, Émilie and Julien Chevallier (15 October 2007). European Carbon Prices and Banking Restrictions: Evidence from Phase I (2005–2007). EconomiX Working Paper 2007-32. EconomiX website, http://economix.u-paris10.fr/en/dt/2007.php?id=53/ [6 June 2008].
148. 111th Congress, 2d Session (2010). American Power Act of 2010, §725. Discussion Draft.
149. Paul Krugman quotes James Hansen's argument without any specific attribution in Krugman, Paul (5 August 2010). Building a Green Economy. *New York Times Magazine*, Online Edition. Dr. Hansen refers to the "perverse effect" of cap-and-trade that renders altruistic actions meaningless in Hansen, James (2009). *Storms of My Grandchildren*, p. 214.
150. Hansen, James (2009). *Storms of My Grandchildren*, p. 213.
151. When the concept of tradable "pollution shares" was initially popularized more than 30 years ago, one of the positive arguments was that environmentalists could "buy up" some of the shares on the market and consequently could reduce the cumulative amount of allowable pollution. I have not heard this argument made in recent decades, presumably because proponents of cap-and-trade mechanisms have noticed that major polluting businesses, such as Exxon or

Con Edison, possess much larger financial resources than environmentalist donors and are therefore much more likely to purchase the shares or allowances.

152. Hansen, James (2009). *Storms of My Grandchildren*, p. 213.
153. Mann, Roberta F. (2009). The Case for the Carbon Tax: How to Overcome Politics and Find Our Green Destiny. *Environmental Law Reporter*, 39(10118). Social Science Research Network website, http://papers.ssrn.com/sol3/papers.cfm?abstract_id=1345181/ [31 March 2009].
154. Hansen, James (2009). *Storms of My Grandchildren*, pp. 211–220, 224.
155. *Id.*, pp. 212–215.
156. *Id.*, pp. 216–217.
157. *Id.*, p. 218.
158. *Id.*
159. Gardner, Timothy and Deborah Charles (10 May 2011). Senate Bill Squeezes Big Oil to Ease Deficit. Thomson Reuters. MSN Money News Center website, http://money.msn.com/business-news/article.aspx?feed=OBR&date=20110510&id=13601956/ [16 June 2011].
160. *Id.*
161. Broder, John M. (13 June 2011). E.P.A. Plans Delay of Rule on Emissions. *New York Times*, Online Edition.
162. Paltsev, Sergey *et al.* (June 2007). Assessment of US Cap-and-Trade Proposals. National Bureau of Economic Research Working Paper 13176.
163. For critiques arguing that cap-and-trade mechanisms are not as effective at eliciting technological innovation as the alternative of direct regulation, see Driesen, David M. (January 2003). Does Emissions Trading Encourage Innovation? *Environmental Law Reporter*, 33, 10094–10108; —— (1998). Free Lunch or Cheap Fix?: The Emissions Trading Idea and the Climate Change Convention. *Boston College Environmental Affairs Law Review*, 26, 1–87.
164. See the scientists' views expressed in an "Open Letter to the Major Economies Forum," prepared for the G8 meeting of July 9, 2009 (European Climate Foundation website, http://www.europeanclimate.org/index.php?option=com_content&task=view&id=52&Itemid=42/ [19 September 2009]).

165. McAllister, Lesley K. (October 2008). The Overallocation Problem in Cap-and-Trade: Moving Toward Stringency. University of San Diego School of Law Research Paper No. 08-076; Reinaud, Julia (October 2008). Climate Policy and Carbon Leakage: Impacts of the European Emissions Trading Scheme on Aluminum. International Energy Agency Information Paper; Burtraw, Dallas (September 2008). Collusion in Auctions for Emission Permits: An Experimental Analysis. Resources for the Future Discussion Paper No. 08-36; Ellerman, A. Denny and Paul L. Joskow (May 2008). The European Union's Emissions Trading System in Perspective. Pew Center on Global Climate Change website, http://www.pewclimate.org/docUploads/EU-ETS-In-Perspective-Report.pdf/ [1 May 2010]; Fujiwara, Noriko and Christian Egenhofer (February 2008). What Lessons Can Be Learned from the EU Emissions Trading Scheme? Centre for European Policy Studies Brief No. 153; Ellerman, A. Denny and Barbara K. Buchner (2007). The European Union Emissions Trading Scheme: Origins, Allocation, and Early Results. *Review of Environmental Economics and Policy*, 1, 66–87; McAllister, Lesley (October 2007). Putting Persuasion Back in the Equation: Compliance in Cap and Trade Programs. University of San Diego Research Paper 07-116; Bleischwitz, Raimund, Katrin Fuhrmann and Elias Huchler (September 2007). The Sustainability Impact of the EU Emissions Trading System on the European Industry. Bruges European Economic Policy Briefing No. 17; Buchner, Barbara, Carlo Carraro and A. Denny Ellerman (September 2006). The Allocation of European Union Allowances: Lessons, Unifying Themes and General Principles. Centre for Economic Policy Research Discussion Paper No. 5483; Betz, Regina and Misato Sato (2006). Emissions Trading: Lessons Learnt from the 1st Phase of the EU ETS and Prospects for the 2nd Phase. *Climate Policy*, 6, 251–259; Egenhofer, Christian *et al.* (July 2006). The EU Emissions Trading Scheme: Taking Stock and Looking Ahead. European Climate Platform Report.
166. Sijm, J.P.M. *et al.* (December 2008). The Impact of the EU ETS on Electricity Prices: Final Report to DG Environment of the European Commission. Energy Research Center of the Netherlands; Point Carbon (March 2008). EU ETS Phase II — The Potential and Scale of Windfall Profits in the Power Sector. Report for WWF; Woerdman,

Edwin, Oscar Couwenberg and Andries Nentjes (November 2007). Energy Prices and Emissions Trading: Windfall Profits from Grandfathering? University of Groningen Faculty of Law Working Paper Series in Law and Economics.

167. Schultz, Abby (13 April 2008). Sizing Up the Utilities, If Carbon Caps Take Hold. *New York Times*, Online Edition; Johnson, Keith (7 April 2008). The $100 Billion Windfall: Why Utilities Love Cap-and-Trade. *Wall Street Journal* website, http://blogs.wsj.com/environmentalcapital/2008/04/07/the-100-billion-windfall-why-utilities-love-cap-and-trade/ [7 April 2008].
168. Hamilton, Katherine *et al.* (20 May 2009). Fortifying the Foundation: State of the Voluntary Carbon Markets. Ecosystem Marketplace and New Carbon Finance Report; Kollmuss, Anja *et al.* (October 2008). A Review of Offset Programs: Trading Systems, Funds, Protocols, Standards and Retailers, Ver. 1.1. Stockholm Environment Institute Report; Monroe, Ian and Lauren Casey (August 2008). US Carbon Markets 101: Making Dollars and Sense Out of Acronym Soup. *Sustainability*, 1(4), 262–268; Wara, Michael W. and David G. Victor (April 2008). A Realistic Policy on International Carbon Offsets. Stanford University Working Paper No. 74; Kollmuss, Anja, Helge Zink and Clifford Polycarp (March 2008). Making Sense of the Voluntary Carbon Market: A Comparison of Carbon Offset Standards. WWF Germany.
169. Parsons, John E., A. Denny Ellerman and Stephan Feilhauer (2009). Designing a US Market for CO_2. *Journal of Applied Corporate Finance*, 21(1), 79–86; Capoor, Karan and Phillipe Ambrosi (May 2009). State and Trends of the Carbon Market 2009. World Bank Report; Larson, Donald F. *et al.* (October 2008). Carbon Markets, Institutions, Policies, and Research. World Bank Policy Research Working Paper No. 4761; Tatsutani, Marika and William A. Pizer (July 2008). Managing Costs in a US Greenhouse Gas Trading Program. Resources for the Future Discussion Paper 08-23.
170. Gans, Joshua S. (October 2007). Do Voluntary Carbon Offsets Work? *Economists' Voice*. Berkeley Electronic Press website, http://www.bepress.com/ev/vol4/iss4/art7/ [24 December 2008]; Revkin, Andrew C. (29 April 2007). Buying Carbon-Neutral. *New York Times*, Online

Edition; —— (29 April 2007). Carbon-Neutral Is Hip, But Is It Green? *New York Times*, Online Edition.

171. Carbon Offsets: Sins of Emission (3 August 2006). *The Economist*, Online Edition.
172. Why Preserve the Rainforest? Marriott website, http://marriott.com/green-brazilian-rainforest.mi/ [27 February 2009].
173. Galst, Liz (2 September 2008). A Balancing Act on Emissions. *New York Times*, Online Edition; Howard, Hilary (30 December 2007). The Green Squadron Keeps Growing. *New York Times*, Online Edition.
174. UNEP's Climate Neutral Network Member PUMA to Spearhead Exciting World Cup Initiative (23 April 2010). UNEP Press Release. Seoul.
175. Gans, Joshua S. (October 2007). Do Voluntary Carbon Offsets Work? *Economists' Voice*. Berkeley Electronic Press website, http://www.bepress.com/ev/vol4/iss4/art7/ [24 December 2008].
176. Krugman, Paul (10 May 2010). Sex & Drugs & the Spill. *New York Times*, Online Edition; Beckenkamp, Martin (January 2009). Environmental Dilemmas Revisited: Structural Consequences from the Angle of Institutional Ergonomics. Max Planck Institute for Research on Collective Goods. MPI Collective Goods Preprint No. 2009/1; Fischbacher, Urs and Simon Gächter (December 2008). Social Preferences, Beliefs, and the Dynamics of Free Riding in Public Good Experiments. CESifo Working Paper No. 2491.
177. Hamilton, Katherine *et al.* (14 June 2010). Building Bridges: State of the Voluntary Carbon Markets 2010. Bloomberg New Energy Finance and Ecosystem Marketplace.
178. Peters-Stanley, Molly *et al.* (2 June 2011). Back to the Future: State of the Voluntary Carbon Markets 2011. Ecosystem Marketplace and Bloomberg New Energy Finance. Ecosystem Marketplace website, http://www.ecosystemmarketplace.com/pages/dynamic/resources.library.page.php?page_id=8351§ion=carbon_market&eod=1/ [16 June 2011].
179. *Id.*, p. iv.
180. US House of Representatives (March 2009). Discussion Draft Summary of the American Clean Energy and Security Act of 2009.
181. 111th Congress, 2d Session (2010). American Power Act of 2010. Discussion Draft.

182. *Id.*, §§731–763, pp. 366–469.
183. *Id.*, §733(c).
184. *Id.*, §733(c)(1)–(7).
185. Congressional Research Service (2010). The American Power Act, Section-by-Section Analysis, §735.
186. Friedman, Thomas L. (8 April 2009). Show Us the Ball. *New York Times*, Online Edition.
187. McKinley, James C., Jr. (9 January 2010). Heavy Rains End Drought for Texas. *New York Times*, Online Edition; Steinhauer, Jennifer (5 June 2008). Governor Declares Drought in California and Warns of Rationing. *New York Times*, Online Edition; Associated Press (30 December 2009). Sierra Snow Below Average, Despite California Storms. *New York Times*, Online Edition; Archibold, Randal C. (10 December 2007). Western States Agree to Water-Sharing Pact. *New York Times*, Online Edition; Goodman, Brenda (16 October 2007). Drought-Stricken South Facing Tough Choices. *New York Times*, Online Edition.
188. Adams, Henry D. *et al*. (28 April 2009). Temperature Sensitivity of Drought-Induced Tree Mortality Portends Increased Regional Die-Off Under Global-Change-Type Drought. *Proceedings of the National Academy of Sciences*, 106(17), 7063–7066.
189. Associated Press (11 June 2010). 16 Dead, Dozens Missing in Arkansas Floods. MSNBC website, http://www.msnbc.msn.com/id/37637416/ns/us_news-life/on/ [11 June 2010]; Harless, William and Liz Robbins (4 May 2010). River Begins to Recede in Tennessee. *New York Times*, Online Edition; Green House (28 April 2010). EPA Says Climate Change Is Causing Storms, Heat Waves. *USA Today* website, http://content.usatoday.com/communities/greenhouse/post/2010/04/epa-report-warns-of-climate-change-dangers/1/ [29 April 2010]; Dewan, Shaila (23 November 2009). In Mississippi Delta, a Promising Summer Washed Away by the Fall. *New York Times*, Online Edition; Brown, Robbie (24 September 2009). Georgia Grapples with Rain's Damage. *New York Times*, Online Edition; Einhorn, Catrin and Malcolm Gay (22 March 2008). Big Storms Continue to Strike Midwest. *New York Times*, Online Edition.
190. Times Topics: California Wildfires (9 January 2010). *New York Times*, Online Edition; Bowman, Quinn (2 September 2009). Scientists: More

Wildfires in West a Consequence of Climate Change. *PBS NewsHour*, Online Edition. http://www.pbs.org/newshour/updates/environment/july-dec09/climatefire_09-02.html/; Steinhauer, Jennifer (17 April 2009). Fearing Wildfires, Utility Plans Power Shutdowns. *New York Times*, Online Edition.

191. Robbins, Jim (18 November 2008). Bark Beetles Kill Millions of Acres of Trees in West. *New York Times*, Online Edition.
192. Navarro, Mireya (23 January 2009). Environment Blamed in Western Tree Deaths. *New York Times*, Online Edition; Fountain, Henry (28 April 2008). The Beetle Factor in a Carbon Calculus. *New York Times*, Online Edition.
193. Eubanks, William S., II (October 2008). The Life-Altering Impacts of Climate Change: The Precipitous Decline of the Northeastern Sugar Maple and the Regional Greenhouse Gas Initiative's Potential Solution. *Pennsylvania State Environmental Law Review*, 17, 101–117; Belluck, Pam (3 March 2007). Warm Winters Upset Rhythms of Maple Sugar. *New York Times*, Online Edition.
194. US Environmental Protection Agency (2010). Climate Change Indicators in the United States, pp. 22–27; Rudolph, John Collins (9 July 2010). The Heat Wave and the Climate Divide. *New York Times*, Online Edition; Goodnough, Abby (6 July 2010). East Swelters in Triple Digits, More to Come. *New York Times*, Online Edition.
195. Associated Press (25 June 2010). In Bridgeport, Conn., 'Deep Sigh of Relief' After Tornado. *New York Times*, Online Edition; Wheaton, Sarah (24 April 2010). Fierce Tornado Causes Deaths Across Mississippi. *New York Times*, Online Edition.
196. Spiegel, Jan Ellen (15 November 2009). Closing Out a Season Farmers Want to Forget. *New York Times*, Online Edition; msnbc.com (12 March 2009). California Counts up Its Warming Costs. MSNBC website, http://www.msnbc.msn.com/id/29656274/.
197. Kramer, Andrew E. (19 July 2010). Russians and Their Crops Wilt Under Heat Wave. *New York Times*, Online Edition; Associated Press (15 July 2010). Bikinis in Moscow: Europe Wilts in Heat Wave. *New York Times*, Online Edition; Wines, Michael (4 April 2010). Spring Harvest of Debt for Parched Farms in Southern China. *New York Times*, Online Edition; Global Risks 2010 (January 2010). World Economic

Forum Report; McMullen, Catherine P. and Jason Jabbour (eds.) (September 2009). UNEP Climate Change Science Compendium 2009; Loayza, Norman *et al.* (June 2009). Natural Disasters and Growth: Going Beyond the Averages. World Bank Policy Research Working Paper 4980; Breshears, David D. *et al.* (18 October 2005). Regional Vegetation Die-Off in Response to Global-Change-Type Drought. *Proceedings of the National Academy of Sciences*, 102(42), 15144–15148.

198. Rosenthal, Elisabeth (22 March 2008). Lofty Pledge to Cut Emissions Comes with Caveat in Norway. *New York Times*, Online Edition.
199. *Id.*
200. Kamprath, Richard (1 November 2009). Norway: A Template for World Energy Policy, Oil and Gas Contracts. Social Science Research Network website, http://ssrn.com/abstract=1577928/ [9 May 2010].
201. *Id.*; Revkin, Andrew C. (22 March 2008). Norway's Green Plans — And Carbon Reality. Dot Earth Blog. *New York Times*, Online Edition.
202. Prasad, Monica (25 March 2008). On Carbon, Tax and Don't Spend. *New York Times*, Online Edition.
203. CDM Executive Board (19 November 2008). Clean Development Mechanism Validation and Verification Manual; Estrada, Manuel, Esteve Corbera and Katrina Brown (May 2008). How Do Regulated and Voluntary Carbon-Offset Schemes Compare? Tyndall Centre for Climate Change Research Working Paper 116; Wara, Michael W. (2008). Measuring the Clean Development Mechanism's Performance and Potential. *UCLA Law Review*, 55, 1759; Flues, Florens, Axel Michaelowa and Katharina Michaelowa (2008). UN Approval of Greenhouse Gas Emission Reduction Projects in Developing Countries: The Political Economy of the CDM Executive Board. Center for Comparative & International Studies Working Paper No. 35.
204. Offset Quality Initiative (2010). Assessing Offset Quality in the Clean Development Mechanism. *Sustainable Development Law & Policy*, 10(2), 25–38; Voigt, Christina (2008). Is the Clean Development Mechanism Sustainable? Some Critical Aspects. *Sustainable Development Law & Policy*, 7(2), 15–21; Gupta, Joyeeta *et al.* (April 2008). An Evaluation of the Contribution of the Clean Development Mechanism to Sustainable Development in Host Countries. Report to the Policy and Operations Evaluation Department of the Netherlands

(IOB), No. 310; Wara, Michael (2006). Measuring the Clean Development Mechanism's Performance and Potential. Stanford Program on Energy and Sustainable Development Working Paper No. 56; Germain, M., A. Magnus and V. van Steenberghe (22 December 2004). Should Developing Countries Participate in the Clean Development Mechanism Under the Kyoto Protocol? The Low-Hanging Fruits and Baseline Issues. Research Paper for the Belgian Federal Science Policy Office.

205. Böhm, Steffen and Siddhartha Dabhi (eds.) (2009). *Upsetting the Offset: The Political Economy of Carbon Markets*. London: Mayfly Books; Chenost, Clément *et al.* (2010). Bringing Forest Carbon Projects to the Market. United Nations Environment Programme.
206. UN-REDD Programme (28 May 2010). MRV in Tanzania. UN-REDD Programme Newsletter, Issue #8; Costenbader, John (ed.) (2009). Legal Frameworks for REDD: Design and Implementation at the National Level. IUCN Environmental Policy and Law Paper No. 77; Fransen, Taryn (June 2009). Enhancing Today's MRV Framework to Meet Tomorrow's Needs: The Role of National Communications and Inventories. World Resources Institute Working Paper.
207. Article 12.5B of the Kyoto Protocol provides that CDM projects must demonstrate "reductions in emissions that are additional to those that would have occurred in the absence of the certified project activity." See Kyoto Protocol to the United Nations Framework Convention on Climate Change, Art. 12.5B (10 December 1997). Reprinted in *International Legal Materials*, 37, 22.
208. CDM/JI Manual for Project Developers and Policy Makers (2007). Japan Ministry of the Environment; Boyd, Emily *et al.* (October 2007). The Clean Development Mechanism: An Assessment of Current Practice and Future Approaches for Policy. Tyndall Centre for Climate Change Research Working Paper No. 114, p. 4.
209. Rosendahl, Knut Einar and Jon Strand (May 2009). Simple Model Frameworks for Explaining Inefficiency of the Clean Development Mechanism. World Bank Policy Research Working Paper No. 4931; Galbraith, Kate (6 October 2008). Double-Dipping in the Offset Marketplace. Green Blog. *New York Times*, Online Edition; Kollmuss, Anja, Helge Zink and Clifford Polycarp (March 2008). Making Sense of

the Voluntary Carbon Market: A Comparison of Carbon Offset Standards. WWF Germany; Estrada, Manuel, Esteve Corbera and Katrina Brown (May 2008). How Do Regulated and Voluntary Carbon-Offset Schemes Compare? Tyndall Centre for Climate Change Research Working Paper 116; Wara, Michael W. and David G. Victor (April 2008). A Realistic Policy on International Carbon Offsets. Stanford University Program on Energy and Sustainable Development Working Paper No. 74; Schneider, Lambert (November 2007). Is the CDM Fulfilling Its Environmental and Sustainable Development Objectives? An Evaluation of the CDM and Options for Improvement. Öko-Institut Report. Berlin.

210. This figure from a UNEP document on CDM projects was reprinted in CDM/JI Manual for Project Developers and Policy Makers (2007). Japan Ministry of the Environment, p. 3, Fig.1-2.

211. Article 12.5B of the Kyoto Protocol provides that CDM projects must demonstrate "reductions in emissions that are additional to those that would have occurred in the absence of the certified project activity." (See Kyoto Protocol to the United Nations Framework Convention on Climate Change, Art. 12.5B (10 December 1997). Reprinted in *International Legal Materials*, 37, 22.) There have been many critical analyses of whether approved CDM projects actually meet the additionality requirement. (See Capoor, Karan and Phillipe Ambrosi (May 2009). State and Trends of the Carbon Market 2009. World Bank Report; Müller, Benito (March 2009). Additionality in the Clean Development Mechanism: Why and What? Oxford Institute for Energy Studies Paper EV 44; Kollmuss, Anja, Helge Zink and Clifford Polycarp (March 2008). Making Sense of the Voluntary Carbon Market: A Comparison of Carbon Offset Standards, pp. 27–32. WWF Germany; Wara, Michael W. and David G. Victor (April 2008). A Realistic Policy on International Carbon Offsets. Stanford University Program on Energy and Sustainable Development Working Paper No. 74.)

212. Rosendahl, Knut Einar and Jon Strand (May 2009). Simple Model Frameworks for Explaining Inefficiency of the Clean Development Mechanism. World Bank Policy Research Working Paper No. 4931; Galbraith, Kate (6 October 2008). Double-Dipping in the Offset Marketplace. Green Blog. *New York Times*, Online Edition; Kollmuss,

Anja, Helge Zink and Clifford Polycarp (March 2008). Making Sense of the Voluntary Carbon Market: A Comparison of Carbon Offset Standards. WWF Germany; Estrada, Manuel, Esteve Corbera and Katrina Brown (May 2008). How Do Regulated and Voluntary Carbon-Offset Schemes Compare? Tyndall Centre for Climate Change Research Working Paper 116; Wara, Michael W. and David G. Victor (April 2008). A Realistic Policy on International Carbon Offsets. Stanford University Program on Energy and Sustainable Development Working Paper No. 74; Schneider, Lambert (November 2007). Is the CDM Fulfilling Its Environmental and Sustainable Development Objectives? An Evaluation of the CDM and Options for Improvement. Öko-Institut Report. Berlin.

213. CDM/JI Manual for Project Developers and Policy Makers (2007). Japan Ministry of the Environment, p. 2, Fig. 1-1.
214. China Approves 51 CDM Projects in Latest Update (10 July 2008). Carbon Market News Service; Bradsher, Keith (6 December 2006). Outsized Profits, and Questions, in Effort to Cut Warming Gases. *New York Times*, Online Edition.
215. Bradsher, Keith (6 May 2010). China's Energy Use Threatens Goals on Warming. *New York Times*, Online Edition; Worldwatch Institute (2007). Coal Use Rises Dramatically Despite Impacts on Climate and Health. *Vital Signs Online*; Vandenbergh, Michael P. (2008). Climate Change: The China Problem. *Southern California Law Review*, 81, 905.
216. Metcalf, Gilbert and David Weisbach (January 2009). The Design of a Carbon Tax. University of Chicago Public Law and Legal Theory Working Paper No. 254; Metcalf, Gilbert E. (2008). Designing a Carbon Tax to Reduce US Greenhouse Gas Emissions. NBER Working Paper 14375; Aldy, Joseph E., Eduardo Ley and Ian W.H. Parry (July 2008). A Tax-Based Approach to Slowing Global Climate Change. Resources for the Future Discussion Paper 08-26; Duff, David G. (May 2003). Tax Policy and Global Warming. University of Toronto Research Paper No. 03-03.
217. Stern, Nicholas (2008). The Economics of Climate Change. *American Economic Review: Papers & Proceedings*, 98(2), 1–37.
218. National Research Council (2010). Hidden Costs of Energy: Unpriced Consequences of Energy Production and Use; Big Oil's Good Deal (11 July 2010). *New York Times*, Online Edition; Krugman, Paul (5 April

2010). Building a Green Economy. *New York Times*, Online Edition; Altemeyer-Bartscher, Martin, Dirk T.G. Rübbelke and Eytan Sheshinski (July 2007). Policies to Internalize Reciprocal International Spillovers. CESifo Working Paper No. 2058; Parry, Ian W.H., Margaret Walls and Winston Harrington (2006). Automobile Externalities and Policies. Resources for the Future Discussion Paper 06-26.

219. Kocieniewski, David (3 July 2010). As Oil Industry Fights a Tax, It Reaps Billions from Subsidies. *New York Times*, Online Edition; Laan, Tara (April 2010). Gaining Traction: The Importance of Transparency in Accelerating the Reform of Fossil-Fuel Subsidies. International Institute for Sustainable Development Global Subsidies Initiative; Broder, John M. (4 March 2010). Lawmakers from Coal States Seek to Delay Emission Limits. *New York Times*, Online Edition; Koplow, Doug (August 2009). Measuring Energy Subsidies Using the Price-Gap Approach: What Does It Leave Out? International Institute for Sustainable Development Trade, Investment & Climate Change Series; Eubanks, William S., II (2009). A Rotten System: Subsidizing Environmental Degradation and Poor Public Health with Our Nation's Tax Dollars. *Stanford Environmental Law Journal*, 28, 213; Hodas, David R. (2007). Ecosystem Subsidies of Fossil Fuels. Widener Law School Legal Studies Research Paper Series No. 08-37.
220. Revkin, Andrew C. (2 August 2010). All's Not Fair When It Comes to Energy Subsidies. *New York Times*, Online Edition; Morales, Alex (29 July 2010). Fossil Fuel Subsidies Are 12 Times Support for Renewables, Study Shows. Bloomberg website, http://www.bloomberg.com/news/2010-07-29/fossil-fuel-subsidies-are-12-times-support-for-renewables-study-shows.html/ [3 August 2010].
221. Zeller, Tom, Jr. (1 April 2010). EPA to Limit Water Pollution from Mining. *New York Times*, Online Edition; Revkin, Andrew C. (23 June 2008). Are Big Oil and Big Coal Climate Criminals? *New York Times*, Online Edition (quoting congressional testimony of Dr. James Hansen).
222. Harvey, L.D. Danny (2009). Reducing Energy Use in the Buildings Sector: Measures, Costs, and Examples. *Energy Efficiency*, 2(2), 139–163.
223. Driesen, David M. and Amy Sinden (2009). The Missing Instrument: Dirty Input Limits. *Harvard Environmental Law Review*, 33, 65.
224. Hansen, James (2009). *Storms of My Grandchildren*.

225. *Id.*, pp. 209–222.
226. *Id.*, p. 209.
227. *Id.*, p. 219.
228. *Id.*, p. 210.
229. *Id.*, p. 213.
230. *Id.*, p. 211.
231. Broder, John M. (4 March 2010). Lawmakers from Coal States Seek to Delay Emission Limits. *New York Times*, Online Edition.
232. Hansen, James (2009). *Storms of My Grandchildren*, pp. 115, 144, 171, 276; Serreze, Mark, Scott B. Luthcke and Konrad Steffen (26 November 2007). Arctic Sea Ice Melt and Shrinking Polar Ice Sheets: Are Observed Changes Exceeding Expectations? Presentation for the American Meteorological Society; Thompson, Andrea (25 March 2008). Ice Shelf on Verge of Collapse: Latest Sign of Global Warming's Impact Shocks Scientists. MSNBC Live Science Blog. MSNBC website, http://www.msnbc.msn.com/id/23797247/ [31 March 2008].
233. Hansen, James (2009). *Storms of My Grandchildren*, p. 172.
234. *Id.*, pp. 185–186.
235. *Id.*, pp. 173–180.
236. *Id.*, p. 209.
237. *Id.*, p. 172.
238. *Id.*, p. 216.
239. *Id.*, p. 210.
240. *Id.*, p. 211.
241. Latin, Howard (1994). Good Warnings, Bad Products, and Cognitive Limitations. *UCLA Law Review*, 41, 1193–1295. Reprinted in A. Bernstein (ed.) (1995). *A Products Liability Anthology*.
242. Hansen, James (2009). *Storms of My Grandchildren*, p. 219.
243. *Id.*, p. 214.
244. *Id.*
245. Parry, Ian W.H. and William A. Pizer (2007). Combating Global Warming: Is Taxation or Cap-and-Trade the Better Strategy for Reducing Greenhouse Emissions? *Regulation*, Fall, 18–22; Rose, Carol M. (2007). Hotspots in the Legislative Climate Change Proposals. University of Arizona Legal Studies Discussion Paper No. 07-36; Ecotaxes: Are Taxes the Best Means to Cut Greenhouse Emissions? (23

April 2007). *The Economist*, Online Edition; Eichner, Thomas and Ruediger Pethig (April 2007). Efficient CO_2 Emissions Control with National Emissions Taxes and International Emissions Trading. CESifo Working Paper No. 1967; Prasad, Monica (25 March 2008). On Carbon, Tax and Don't Spend. *New York Times*, Online Edition.

246. Gardner, Timothy and Deborah Charles (10 May 2011). Senate Bill Squeezes Big Oil to Ease Deficit. Thomson Reuters and MSN news story. On file with the author.
247. Bradsher, Keith (1 September 2011). China Benefits as U.S. Solar Industry Withers. *New York Times*, Online Edition.
248. Feebates: A Complementary Strategy for Reducing GHG from Vehicles (May 2010). UC Davis Institute of Transportation Studies. Research Finding No. 1; Fischer, Carolyn and Alan K. Fox (May 2009). Combining Rebates with Carbon Taxes: Optimal Strategies for Coping with Emissions Leakage and Tax Interactions. Resources for the Future Discussion Paper 09-12.
249. Latin, Howard (1985). Ideal Versus Real Regulatory Efficiency: Implementation of Uniform Standards and "Fine-Tuning" Regulatory Reforms. *Stanford Law Review*, 37, 1267–1332. Reprinted in R. Fischman, M. Lipeles and M. Squillace (eds.) (1996). *An Environmental Law Anthology*. Also reprinted in *Land Use & Environment Law Review* (1987).
250. Castro, Paula and Axel Michaelowa (2008). Climate Strategies Report: Empirical Analysis of Performance of CDM Projects; Rosenthal, Elisabeth (23 April 2008). Europe Turns Back to Coal, Raising Climate Fears. *New York Times*, Online Edition; Block, Ben (10 March 2008). EU Behind in Meeting Key Environmental Targets. Worldwatch Institute website, http://www.worldwatch.org/node/5642/ [13 March 2008]; Revkin, Andrew (21 January 2008). The Climate Challenge. Same as It Ever Was? *New York Times*, Online Edition; Mufson, Steven (9 April 2007). Europe's Problems Color US Plans to Curb Carbon Gases. *Washington Post*, Online Edition.
251. Point Carbon Advisory Services (March 2008). EU ETS Phase II — The Potential and Scale of Windfall Profits in the Power Sector. A Report for WWF; Egenhofer, Christian *et al.* (July 2006). The EU Emissions Trading Scheme: Taking Stock and Looking Ahead. European Climate Platform Report; Johnston, Angus (May 2006). Free Allocation of

Allowances Under the EU Emissions Trading System: Legal Issues. *Climate Policy,* 6, 115; Herro, Alana (17 July 2006). Kyoto: Impossible Goal or Economic Opportunity? Worldwatch Institute website, http://www.worldwatch.org/node/4362/ [27 September 2006].

252. Winter, Gerd (2010). The Climate Is No Commodity: Taking Stock of the Emissions Trading System. *Journal of Environmental Law,* 22(1), 1–25; Kanter, James (2 April 2009). EU Carbon Trading System Shows Signs of Working. *New York Times,* Online Edition; Johnson, Keith (7 April 2008). The $100 Billion Windfall: Why Utilities Love Cap-and-Trade. Environmental Capital Blog. *Wall Street Journal,* Online Edition.
253. Kanter, James (29 August 2010). Cap-and-Trade Is Beginning to Raise Some Concerns. *New York Times,* Online Edition.
254. United Nations Framework Convention on Climate Change (1992). UN Doc. A/AC.237/18. Reprinted in *International Legal Materials,* 31, 849.
255. Kyoto Protocol to the United Nations Framework Convention on Climate Change (10 December 1997). UN Doc. FCCC/CP/1997/L.7/Add.1, Art. 3.1 & Annex B. Reprinted in *International Legal Materials,* 37, 22.
256. The complete text of the Copenhagen Accord can be downloaded from the UNFCCC website at the following URL: http://unfccc.int/files/meetings/cop_15/application/pdf/cop15_cph_auv.pdf/.
257. Mouawad, Jad and Andrew C. Revkin (14 October 2009). Saudis Seek Payments for Any Drop in Oil Revenues. *New York Times,* Online Edition.
258. Bodansky, Daniel (16 February 2010). The Copenhagen Climate Change Conference: A Post-Mortem. University of Georgia School of Law Working Paper.
259. Dubash, Navroz K. (26 December 2009). Copenhagen: Climate of Mistrust. *Economic & Political Weekly,* 44(52), 8–11.
260. Fowler, Rob (20 December 2009). An Initial Assessment of the Copenhagen Outcomes. University of South Australia Law School.
261. Thome, Wolfgang H. (23 December 2009). National Interests Supersede Global Obligations: How Copenhagen Betrayed Africa. eTurboNews website, http://www.eturbonews.com/13448/how-copenhagen-betrayed-africa/ [24 December 2009].
262. Fry, Ian (18 March 2010). Moving Beyond Copenhagen — A Small Island State's Perspective. Guest Article No. 37. Climate-L website, http://climate-l.iisd.org/guest-articles/moving-beyond-copenhagen-

%E2%80%93-a-small-island-state%E2%80%99s-perspective/ [27 May 2010].

263. Dubash, Navroz K. (26 December 2009). Copenhagen: Climate of Mistrust. *Economic & Political Weekly*, 44(52), 11.
264. Müller, Benito (November 2009). The Time Is Right! Devolution of Funding Decisions to Designated National/Regional Climate Change Funding Entities. Oxford Institute for Energy Studies Energy and Environment Comment.
265. Zeller, Tom, Jr. (14 December 2009). Week of Posturing at Copenhagen Climate Talks. *New York Times*, Online Edition; Zeller, Tom, Jr. (5 December 2009). Negotiators at Climate Talks Face Deep Fault Lines. *New York Times*, Online Edition; Dubash, Navroz K. (24 November 2009). Opinion: Other Countries Have Climate Politics, Too. Worldwatch Institute website, http://www.worldwatch.org/node/6324/ [4 December 2009]; —— (2009). Climate Politics in India: How Can the Industrialized World Bridge the Trust Deficit? In David Michel and Amit Pandya (eds.). Indian Climate Policy: Choices and Challenges. Stimson Center Regional Voices Report.
266. Sanwal, Mukul (3 April 2010). The Post-Copenhagen Agenda: Leadership, Climate Justice and the Carbon Budget. Carbon Positive; Dubash, Navroz K. (26 December 2009). Copenhagen: Climate of Mistrust. *Economic & Political Weekly*, 44(52), 8–9.
267. UN Economic and Social Council, Economic Commission for Africa and African Union Commission (5 March 2008). Climate Change: African Perspectives for a Post-2012 Agreement; Srinivasan, Ancha *et al.* (2008). The Climate Regime Beyond 2012: Reconciling Asian Developmental Priorities and Global Climate Interests. Institute for Global Environmental Strategies; Stevenson, Andrew and Christine Loh (21 December 2007). Bali's Hope and Asia's Future. Chinadialogue website, http://www.chinadialogue.net/article/show/single/en/1594-Bali-s-hopes-and-Asia-s-future/ [22 January 2008]; Buys, Piet *et al.* (August 2007). Country Stakes in Climate Change Negotiations: Two Dimensions of Vulnerability. World Bank Policy Research Working Paper 4300; Pizer, William A. (2007). A US Perspective on Future Climate Regimes. Resources for the Future Discussion Paper; Bitter Divisions Exposed at Climate Talks (14 December 2008). *New York Times*, Online Edition.

268. United Nations Framework Convention on Climate Change (1992). UN Doc. A/AC.237/18. Concluded at Rio de Janeiro on 29 May 1992. Reprinted in *International Legal Materials*, 31, 849.
269. Hansen, James (2009). *Storms of My Grandchildren*, pp. 189–190.
270. Associated Press (9 December 2007). US 'Not Ready' to Commit to Bali. Reprinted in *New York Times*, Online Edition.
271. Government of India (2008). National Action Plan on Climate Change. Prime Minister's Council on Climate Change.
272. *Id.*, p. 3.
273. Dubash, Navroz K. (26 December 2009). Copenhagen: Climate of Mistrust. *Economic & Political Weekly*, 44(52), 8.
274. Srinivasan, Ancha *et al.* (2008). The Climate Regime Beyond 2012: Reconciling Asian Developmental Priorities and Global Climate Interests. Institute for Global Environmental Strategies; Dernbach, John C. (27 January 2008). Energy Efficiency and Conservation as Ethical Responsibilities: Suggestions for IPCC Working Group III. Social Science Research Network website, http://papers.ssrn.com/sol3/papers.cfm?abstract_id=1089423/ [13 June 2008]; Müller, Benito, Niklas Höhne and Christian Ellermann (October 2007). Differentiating (Historic) Responsibilities for Climate Change. Oxford Institute for Energy Studies Summary Report; Gosseries, Axel (2004). Historical Emissions and Free Riding. *Ethical Perspectives*, 11(1), 36; Developed Nations Asked to Take Lead in Climate Conservation (17 June 2007). *Hindustan Times*.
275. United Nations Millennium Declaration (September 2000); International Bank for Reconstruction and Development and World Bank (2006). Global Monitoring Report on the Millennium Development Goals: Strengthening Mutual Accountability — Aid, Trade and Governance; World Bank (2006). The Road to 2050: Sustainable Development for the 21st Century.
276. Ban Urges Focus on Development and Green Growth Ahead of G20 Summit (21 June 2010). UN News Service.
277. *Id.*
278. Agence France-Presse (8 December 2007). Developing World Must Be Able to Lift Emissions: Nobel Winner.

279. Revkin, Andrew C. (6 April 2008). A Shift in the Debate over Global Warming. *New York Times,* Online Edition; Sachs, Jeffrey D. (March 2008). Climate Change After Bali. *Scientific American,* 298(3), 33–34.
280. Baker, Peter (9 July 2009). Poorer Nations Reject a Target on Emission Cut. *New York Times,* Online Edition.
281. Revkin, Andrew C. (2 September 2010). China Sustains Blunt 'You First' Message on CO_2. *New York Times,* Online Edition.
282. Dubash, Navroz K. (26 December 2009). Copenhagen: Climate of Mistrust. *Economic & Political Weekly,* 44(52), 9; Baker, Peter (9 July 2009). Poorer Nations Reject a Target on Emission Cut. *New York Times,* Online Edition.
283. Müller, Benito and Harald Winkler (February 2008). One Step Forward, Two Steps Back? The Governance of the World Bank Climate Investment Funds. Oxford Energy & Environment Comment; Srinivasan, Ancha *et al.* (2008). The Climate Regime Beyond 2012: Reconciling Asian Developmental Priorities and Global Climate Interests. Institute for Global Environmental Strategies (many references to meeting the UN Millennium Development Goals despite climate change problems); Buys, Piet *et al.* (August 2007). Country Stakes in Climate Change Negotiations: Two Dimensions of Vulnerability. World Bank Policy Research Working Paper 4300; Working Group on Climate and Development (2004). Africa — Up in Smoke? Previously available at http://www.neweconomics.org/publications/africa-smoke/ [23 July 2008]. On file with the author.
284. India Plans Non-Targets Climate Action (11 February 2008). Carbon Positive website, http://www.carbonpositive.net/viewarticle.aspx?articleID=989/ [10 March 2008].
285. Cover Story — A Joke on the World (11 December 2007). *Down to Earth: Science and Environment Online.*
286. Hunter, David, James Salzman and Durwood Zaelke (2007). *International Environmental Law & Policy, 3rd Edition,* pp. 516–521. New York: Foundation Press; Anton, Donald K. *et al.* (2007). *International Environmental Law: Cases, Materials, Problems,* pp. 618–621. Newark: Matthew Bender & Co.

287. United Nations Framework Convention on Climate Change (1992). UN Doc. A/AC.237/18. Concluded at Rio de Janeiro on 29 May 1992. Reprinted in *International Legal Materials*, 31, 849.
288. Kyoto Protocol to the United Nations Framework Convention on Climate Change (10 December 1997). UN Doc. FCCC/CP/1997/L.7/Add.1. Reprinted in *International Legal Materials*, 37, 22.
289. *Id.*, Articles 2 & 3.
290. Nations Have Mixed Reactions to Copenhagen (24 December 2009). *Tiempo Climate Newswatch*.
291. Zhang, Zhong Xiang (July 2010). Assessing China's Energy Conservation and Carbon Intensity: How Will the Future Differ from the Past? In Ross Garnaut, Jane Golley and Ligang Song (eds.). *China: The Next Twenty Years of Reform and Development*. Canberra: Australian National University E-Press and Brookings Institution Press.
292. Rapp, Tobias, Christian Schwägerl and Gerald Traufetter (5 May 2010). How China and India Sabotaged the UN Climate Summit. *Spiegel Online International* website, http://www.spiegel.de/international/world/0,1518,692861,00.html/ [4 September 2010].
293. Lee, Roy (26 May 2010). A Way Forward from the Current Impasse in Climate Negotiation. Guest Article No. 41. Climate-L website, http://climate-l.iisd.org/guest-articles/a-way-forward-from-the-current-impasse-in-climate-negotiation/ [27 May 2010].
294. Wong, Edward and John M. Broder (27 November 2009). US and China to Go to Talks with Emissions Targets. *New York Times*, Online Edition. This article makes the same mistake as Roy Lee did in claiming that increasing the intensity of economic productivity is equivalent to "reducing" GHG emissions.
295. Revkin, Andrew C. and Tom Zeller, Jr. (10 December 2009). US Negotiator Dismisses Reparations for Climate. *New York Times*, Online Edition.
296. Antholis, William (20 July 2009). Opinion: India and Climate Change. *Wall Street Journal*, Online Edition.
297. Mackey, Brendan and Parvez Hassan (19 January 2009). Contraction & Convergence — A Framework for Ethically Closing the Mitigation Implementation Gap. Climate Ethics Blog. Rock Ethics Institute website, http://rockblogs.psu.edu/climate/about-climate-ethics.html/ [24 November 2009]; Global Commons Institute (2008). Carbon

. Countdown: The Campaign for Contraction & Convergence. Global Commons Institute website, http://www.gci.org.uk/kite/Carbon_Countdown.pdf/ [5 December 2008].

298. Brown, Donald A. (July 2002). The US Performance in Achieving Its 1992 Earth Summit Global Warming Commitments. *Environmental Law Reporter*, 32, 10741; Manne, Alan S. and Richard G. Richels (October 2001). US Rejection of the Kyoto Protocol: The Impact on Compliance Costs and CO_2 Emissions. American Enterprise Institute and Brookings Joint Center for Regulatory Studies Working Paper 01-12.
299. Broder, John M. (3 June 2008). Senate Opens Debate on Politically Risky Bill Addressing Global Warming. *New York Times*, Online Edition; Harrabin, Roger (18 January 2008). EU Climate Policy 'Too Negative'. BBC News website, http://news.bbc.co.uk/2/hi/science/nature/7194250.stm/ [31 January 2008].
300. Jordan-Korte, Katrin and Stormy Mildner (June 2008). Climate Protection and Border Tax Adjustment: Economic Rationale and Political Pitfalls of Current U.S. Cap-and-Trade Proposals. FACET Analysis No. 1.
301. Moore, Michael O. (July 2010). Implementing Carbon Tariffs: A Fool's Errand? World Bank Policy Research Working Paper No. 5359.
302. Zhang, Zhong Xiang (2010). The US Proposed Carbon Tariffs, WTO Scrutiny and China's Responses. Fondazione Eni Enrico Mattei Note di Lavoro Series No. 34.2010.
303. Reuters (28 November 2010). Modest Climate Change Steps Are Goal of Meeting in Mexico. *New York Times*, Online Edition.
304. UNEP (11 December 2010). UN Climate Change Conference in Cancun Delivers Balanced Package of Decisions, Restores Faith in Multilateral Process. UNEP News Service. UNEP website, http://www.unep.org/Documents.Multilingual/Default.asp?DocumentID=653&ArticleID=6866&l=en/.
305. Broder, John M. (11 December 2010). Climate Talks End with Modest Deal on Emissions. *New York Times*, Online Edition.
306. UNEP (11 December 2010). UN Climate Change Conference in Cancun Delivers Balanced Package of Decisions, Restores Faith in Multilateral Process. UNEP News Service. UNEP website, http://www.unep.org/Documents.Multilingual/Default.asp?DocumentID=653&ArticleID=6866&l=en/.

307. *Id.*
308. *Id.*
309. UN Framework Convention on Climate Change (2007). Bali Road Map. UNFCCC website, http://unfccc.int/key_documents/bali_road_map/items/6447.php/.
310. Fuller, Thomas and Andrew C. Revkin (16 December 2007). Nations Agree on Steps to Revive Climate Treaty. *New York Times*, Online Edition.
311. Revkin, Andrew C. and John M. Broder (20 December 2009). A Grudging Accord in Climate Talks. *New York Times*, Online Edition.
312. Torney, Diarmuid and Annika Greup (2010). Editorial Introduction: New Directions in Climate Change Politics. *St Antony's International Review*, 5(2), 5–15.
313. UN Environment Programme (11 December 2010). UN Climate Change Conference in Cancun Delivers Balanced Package of Decisions, Restores Faith in Multilateral Process.
314. La Vina, Antonio G.M., Lawrence Ang and JoAnne Dulce (2010). The Cancun Agreements: Do They Advance Global Cooperation on Climate Change? Foundation for International Environmental Law and Development Working Paper. FIELD website, http://www.field.org.uk/.
315. Broder, John M. (11 December 2010). Climate Talks End with Modest Deal on Emissions. *New York Times*, Online Edition.
316. UN Environment Programme (11 December 2010). UN Climate Change Conference in Cancun Delivers Balanced Package of Decisions, Restores Faith in Multilateral Process.
317. *Id.*
318. IISD Reporting Services (13 December 2011). Earth Negotiations Bulletin: Summary of the Durban Climate Change Conference (28 November–11 December 2011), p. 1 [hereafter cited as IISD Summary].
319. *Id.*
320. *Id.*, pp. 1–31.
321. Andrews, Paige *et al.* (January 2012). COP-17 De-briefing: Enhancements, Decisions, and the Durban Package. Climatico Analysis website, http://www.climaticoanalysis.org/reports/; Sterk, Wolfgang *et al.* (December 2011). On the Road Again: Progressive Countries Score a Realpolitik Victory in Durban While the Real Climate Continues to Heat Up. Wuppertal Institute Working Paper; Honig, Shira (11 December 2011).

A Surprise Ending for Durban (Almost). Climatico Analysis website, http://www.climaticoanalysis.org/post/a-surprise-ending-for-durban-almost/.

322. UNFCCC COP 17 (December 2011). Establishment of an Ad Hoc Working Group on the Durban Platform for Enhanced Action. Draft Decision -/CP.17. UNFCCC website, http://unfccc.int/files/meetings/durban_nov_2011/decisions/application/pdf/cop17_durbanplatform.pdf/.
323. Maxwell, Simon (12 January 2012). Reflections on the Durban Outcomes. Climate and Development Knowledge Network website, http://cdkn.org/2012/01/reflections-on-the-durban-outcomes/?loclang = en_gb/; Peszko, Grzegorz (21 December 2011). Durban Platform: Breakthrough or Procrastination?, p. 1. European Bank for Reconstruction and Development Blog.
324. Editorial Board (14 December 2011). Are Big International Conferences Useless? *Washington Post*, Online Edition; Boyle, Jessica (December 2011). Assessing the Outcomes of COP 17 — In Pursuit of a Binding Climate Agreement: Negotiators Expand the Mitigation Tent But Reinforce the Ambition Gap. International Institute for Sustainable Development website, http://www.iisd.org/publications/pub.aspx?id = 1536/.
325. IISD Summary, pp. 26–27.
326. EUROPA (12 December 2011). Durban Conference Delivers Breakthrough for Climate. Think to Sustain: A Market Space for Ideas website, http://www.thinktosustain.com/infocusdetails.aspx?id = 260/; Lee, Bernice (11 December 2011). Expert Comment — Durban Climate Deal: Staying Alive to Fight Another Day. Chatham House website, http://www.chathamhouse.org/media/comment/view/180495/.
327. IISD Summary, pp. 21–23, 25.
328. UNFCCC COP 17 (December 2011). Establishment of an Ad Hoc Working Group on the Durban Platform for Enhanced Action. Draft Decision -/CP.17. UNFCCC website, http://unfccc.int/files/meetings/durban_nov_2011/decisions/application/pdf/cop17_durbanplatform.pdf/; IISD Summary, pp. 25–26.
329. IISD Summary, pp. 15, 20.
330. Sterk, Wolfgang *et al.* (December 2011). On the Road Again: Progressive Countries Score a Realpolitik Victory in Durban While the Real Climate Continues to Heat Up. Wuppertal Institute Working Paper, p. 9.

331. UNFCCC COP 17 (December 2011). Establishment of an Ad Hoc Working Group on the Durban Platform for Enhanced Action. Draft Decision -/CP.17, paragraph 6. UNFCCC website, http://unfccc.int/files/meetings/durban_nov_2011/decisions/application/pdf/cop17_durbanplatform.pdf/.
332. *Id.*, paragraph 7.
333. Sterk, Wolfgang *et al.* (December 2011). On the Road Again: Progressive Countries Score a Realpolitik Victory in Durban While the Real Climate Continues to Heat Up. Wuppertal Institute Working Paper, p. 6.
334. IISD Summary, p. 17.
335. *Id.*, p. 26.
336. *Id.*, p. 27.
337. Robinson, Eugene (12 December 2011). Reason to Smile About the Durban Conference. *Washington Post,* Online Edition.
338. *Id.*
339. Sterk, Wolfgang *et al.* (December 2011). On the Road Again: Progressive Countries Score a Realpolitik Victory in Durban While the Real Climate Continues to Heat Up. Wuppertal Institute Working Paper.
340. Moomaw, William R. and Mihaela Papa (13 December 2011). Two Decades of Failed Diplomacy — Time for a Change. Dot Earth Blog. *New York Times,* Online Edition.
341. Flavin, Christopher (23 December 2009). Escape from Copenhagen. Worldwatch Institute.
342. UN Economic and Social Council, Economic Commission for Africa and African Union Commission (5 March 2008). Climate Change: African Perspectives for a Post-2012 Agreement; Srinivasan, Ancha *et al.* (2008). The Climate Regime Beyond 2012: Reconciling Asian Developmental Priorities and Global Climate Interests. Institute for Global Environmental Strategies; Stevenson, Andrew and Christine Loh (21 December 2007). Bali's Hope and Asia's Future. Chinadialogue website, http://www.chinadialogue.net/article/show/single/en/594-Bali-s-hopes-and-Asia-s-future/ [22 January 2008]; Buys, Piet *et al.* (August 2007). Country Stakes in Climate Change Negotiations: Two Dimensions of Vulnerability. World Bank Policy Research Working Paper 4300; Pizer, William A. (2007). A US Perspective on Future Climate Regimes. Resources for the Future Discussion Paper; Fuller, Thomas and Elisabeth

Rosenthal (14 December 2007). Gore Joins Chorus Chiding US at Climate Talks. *New York Times,* Online Edition.

343. Hansen, James (2009). *Storms of My Grandchildren*, pp. 115, 144, 176.
344. Bradsher, Keith (17 June 2010). Security Tops Environment in China Energy Plan. *New York Times*, Online Edition; —— (6 May 2010). China's Energy Use Threatens Goals on Warming. *New York Times*, Online Edition; World Bank Supports Modernization of Old, Polluting Coal-Fired Power Plants in India to Lower Carbon Emissions (18 June 2009). World Bank Press Release No. 2009/417/SAR; Prins, Gwyn (4 April 2008). The Road from Kyoto. *The Guardian*; Wheeler, David and Kevin Ummel (December 2007). Another Inconvenient Truth: A Carbon-Intensive South Faces Environmental Disaster No Matter What the North Does. Center for Global Development Working Paper 134; Associated Press (7 December 2007). China, US Face Off on Climate Policies. MSNBC website, http://www.msnbc.msn.com/id/22148697/ [12 December 2007]; Mouawad, Jad (7 November 2007). Cuts Urged in China's and India's Energy Growth. *New York Times*, Online Edition.
345. Vandenbergh, Michael P. (2008). Climate Change: The China Problem. *Southern California Law Review*, 81, 905; Zhang, Zhong Xiang (2007). China, the United States and Technology Cooperation on Climate Control. *Environmental Science and Policy*, 10(7–8), 622–628; —— (2000). Can China Afford to Commit Itself to an Emissions Cap? An Economic and Political Analysis. *Energy Economics*, 22, 587–614; Sunstein, Cass R. (7 August 2007). The Complex Climate Change Incentives of China and the United States. University of Chicago Law & Economics Olin Working Paper No. 352; India Plans Non-Targets Climate Action (11 February 2008). Carbon Positive website, http://www.carbonpositive.net/viewarticle.aspx?articleID = 989/ [10 March 2008]; Narain, Sunita (5 August 2008). The Mean World of Climate Change. *Down to Earth: Science and Environment Online.*
346. Kahn, Joseph and Mark Landler (27 December 2007). China Grabs West's Smoke-Spewing Factories. *New York Times*, Online Edition; Friedman, Thomas L. (4 November 2007). No, No, No, Don't Follow Us. *New York Times*, Online Edition.

347. Stolberg, Sheryl Gay (17 April 2008). Bush Sets Greenhouse Gas Emissions Goal. *New York Times*, Online Edition; Barringer, Felicity and William Yardley (4 April 2007). Bush Splits with Congress and States on Emissions. *New York Times*, Online Edition.
348. UNFCCC (2010). Technology Executive Committee, Function (d). UNFCCC website, http://unfccc.int/ttclear/jsp/TEC.jsp/.
349. UNFCCC (2010). Climate Technology Centre and Network, Function (a)(i). UNFCCC website, http://unfccc.int/ttclear/jsp/CTCN.jsp/.
350. IISD Summary, p. 19.
351. UNFCCC (2012). Full Text of the Convention, Article 4(c). UNFCCC website, http://unfccc.int/essential_background/convention/background/items/1362.php/.
352. Moomaw, William R. and Mihaela Papa (13 December 2011). Two Decades of Failed Diplomacy — Time for a Change. Dot Earth Blog. *New York Times*, Online Edition.
353. Hansen, James (2009). *Storms of My Grandchildren*, p. 219.
354. *Id.*, p. 220.
355. Revkin, Andrew C. (2 September 2010). China Sustains Blunt 'You First' Message on CO_2. *New York Times*, Online Edition; Cooper, Helene and John M. Broder (19 December 2009). Obama Presses China on Rules for Monitoring Emissions Cuts. *New York Times*, Online Edition.
356. Zhang, Zhong Xiang (February 2010, revised March 2010). Copenhagen and Beyond: Reflections on China's Stances and Response. Speech at the International Workshop on Climate Change Policies. Madrid, Spain; Associated Press (11 December 2009). China, US Argue as Climate Draft Takes Shape. MSNBC website, http://www.msnbc.msn.com/id/34378495/ [12 December 2009].
357. Reuters (26 May 2010). Mainland Official All But Dashes Hopes of Climate Deal This Year. *South China Morning Post*; Thome, Wolfgang H. (23 December 2009). National Interests Supersede Global Obligations: How Copenhagen Betrayed Africa. eTurboNews website, http://www.eturbonews.com/13448/how-copenhagen-betrayed-africa/ [24 December 2009]; Heilprin, John (22 September 2009). China, India Focus of UN Climate Summit. *Time*, Online Edition. http://www.time.com/time/world/article/0,8599,1925303,00.html/ [22 September 2009].

358. Revkin, Andrew C. (2 September 2010). China Sustains Blunt 'You First' Message on CO_2. *New York Times*, Online Edition; Bradsher, Keith (17 June 2010). Security Tops Environment in China Energy Plan. *New York Times*, Online Edition; —— (6 May 2010). China's Energy Use Threatens Goals on Warming. *New York Times*, Online Edition.
359. Bradsher, Keith (31 January 2010). China Leading Global Race to Make Clean Energy. *New York Times*, Online Edition; Bradsher, Keith (11 May 2009). China Emerges as a Leader in Cleaner Coal Technology. *New York Times*, Online Edition; Ma, Hengyun, Les Oxley and John Gibson (April 2009). China's Energy Situation and Its Implications in the New Millennium. Motu Economic and Public Policy Research Working Paper No. 09-04.
360. Bradsher, Keith (6 May 2010). China's Energy Use Threatens Goals on Warming. *New York Times*, Online Edition.
361. Revkin, Andrew C. (4 September 2010). Views on China's Role in the Greenhouse. *New York Times*, Online Edition.
362. Krugman, Paul (18 May 2009). The Perfect, the Good, the Planet. *New York Times*, Online Edition.
363. Another Failure on Climate Change (11 June 2008). *New York Times*, Online Edition.
364. Coglianese, Cary and Jocelyn D'Ambrosio (2008). Policymaking Under Pressure: The Perils of Incremental Responses to Climate Change. University of Pennsylvania Law School Research Paper No. 08-30.
365. Crabb, Joseph M. and Daniel K.N. Johnson (May 2007). Fueling the Innovative Process: Oil Prices and Induced Innovation in Automotive Energy-Efficient Technology. Colorado College Working Paper 2007-04; Garrett, Jerry (20 April 2008). A Redesign Waiting for Diesel. *New York Times*, Online Edition; Motavalli, Jim (July–August 2004). Getting There: A Guide to Planet-Friendly Cars. *E — The Environmental Magazine*.
366. Examples of the five alternative vehicle technologies discussed were previously available at http://chevy.nytimes.com/.
367. *Id.*
368. Niedermeyer, Edward (29 July 2010). GM's Electric Lemon. *New York Times*, Online Edition.

369. The 2010 Paris Motor Show featured numerous all-electric and low-GHG vehicles including both future designs and current models, but American car manufacturers played only a small role in this show. See Jolly, David (29 September 2010). Paris Auto Show Focuses on Green Cars as Automakers Focus on China. *New York Times*, Online Edition.
370. The American auto manufacturers have already challenged in court California and federal fuel-efficiency act requirements that say they must cut GHG emissions by 30% by 2020, and they eventually resigned themselves to complying with this standard. (See Sperling, Daniel and James S. Cannon (eds.) (2010). Climate and Transportation Solutions: Findings from the 2009 Asilomar Conference on Transportation and Energy Policy. Institute of Transportation Studies, University of California at Davis, California.) The companies were using the interesting argument that they had invested heavily in large gas-guzzling vehicles that have now become unpopular because of rapidly increasing gas prices, and as a result they could not afford to build future vehicles efficient enough to meet the regulatory standards. (See Big Carmakers Say Fuel Rule Plan Too Strict (1 July 2008). *New York Times*, Online Edition.)
371. The 2008 GM advertisement was previously available at http://chevy.nytimes.com/.
372. Sperling, Daniel and Deborah Gordon (2009). *Two Billion Cars: Driving Toward Sustainability*. Oxford: Oxford University Press.
373. See endnote 333.
374. US Department of Energy (31 January 2008). Clean Coal Technology & the President's Clean Coal Power Initiative. US Department of Energy website, http://www.fossil.energy.gov/programs/powersystems/cleancoal/ [31 January 2008].
375. World Coal Institute (2008). Coal Meeting the Climate Challenge: Technology to Reduce Greenhouse Gas Emissions; US Department of Energy (31 January 2008). Hydrogen & Clean Fuels Research. US Department of Energy website, http://www.fossil.energy.gov/programs/fuels/ [31 January 2008]; US Department of Energy (16 November 2007). Energy Efficiency and Renewable Energy. Alternative Fuels and Advanced Vehicles Data Center website, http://www.afdc.energy.gov/afdc/ [17 November 2007]; Russell, James (9 November

2007). US Government Dumping $100 Million into Filthy Fuels Project. Worldwatch Institute website, http://www.worldwatch.org/node/5487/ [16 November 2007]; Andrews, Edmund L. (29 May 2007). Lawmakers Push for Big Subsidies for Coal Process. *New York Times*, Online Edition.

376. Wald, Matthew L. (5 February 2008). Utilities Turn from Coal to Gas, Raising Risk of Price Increase. *New York Times*, Online Edition.
377. Fischedick, Manfred *et al.* (2007). CO_2-Capture and Geological Storage as a Climate Policy Option. Wuppertal Institute Special Report 35e; Vallentin, Daniel (April 2007). Inducing the International Diffusion of Carbon Capture and Storage Technologies in the Power Sector. Wuppertal Paper No. 162; US Department of Energy (2007). Carbon Sequestration Technology Roadmap and Program Plan; de Coninck, Heleen (June 2008). The International Race for Carbon Capture and Storage: And the Winner Is ...? Forum for Atlantic Climate and Energy Talks (FACET) Commentary No. 12.
378. Mann, Roberta (2007). Another Day Older and Deeper in Debt: How Tax Incentives Encourage Burning Coal and the Consequences for Global Warming. *Pacific McGeorge Global Business & Development Law Journal*, 20, 111–141; Andrews, Edmund L. (29 May 2007). Lawmakers Push for Big Subsidies for Coal Process. *New York Times*, Online Edition.
379. Nace, Ted (20 March 2008). The World's Dumbest Project: Tata Ultra Mega. Grist website, http://www.grist.org/article/worlds-dumbest-project-tata-ultra-mega/ [20 July 2008].
380. *Id.*; Wheeler, David (13 March 2008). Tata Ultra Mega Mistake: The IFC Should Not Get Burned by Coal. CARMA website, http://carma.org/blog/tata-ultra-mega-mistake-the-ifc-should-not-get-burned-by-coal/ [20 July 2008].
381. Coglianese, Cary and Jocelyn D'Ambrosio (2008). Policymaking Under Pressure: The Perils of Incremental Responses to Climate Change. *Connecticut Law Review*, 40, 1411.
382. Reitze, Arnold W., Jr. (July–August 2007). Should the Clean Air Act Be Used to Turn Petroleum Addicts into Alcoholics? *Environmental Forum*, pp. 50–60; Krauss, Clifford (18 December 2007). As Ethanol Takes Its First Steps, Congress Proposes a Giant Leap. *New York Times*, Online

Edition; Widenoja, Raya (October 2007). Destination Iowa: Getting to a Sustainable Biofuels Future. Sierra Club and Worldwatch Institute Report; Greene, Nathanael (December 2004). Growing Energy: How Biofuels Can Help End America's Oil Dependence. Natural Resources Defense Council Report.

383. Another Problem with Biofuels? (12 March 2008). *Time* website, http://www.time.com/time/health/article/0,8599,1721693,00.html/ [14 March 2008]; Rosenthal, Elisabeth (8 February 2008). Biofuels Deemed a Greenhouse Threat. *New York Times*, Online Edition; Rajagopal, Deepak and David Zilberman (September 2007). Review of Environmental, Economic and Policy Aspects of Biofuels. World Bank Policy Research Working Paper No. 4341; Martin, Andrew (18 December 2007). Food and Fuel Compete for Land. *New York Times*, Online Edition; Mongoven, Bart (13 September 2007). The Biofuel Backlash. STRATFOR Public Policy Intelligence Report. STRATFOR Global Intelligence website, http://www.stratfor.com/biofuel_backlash?fn=3013236997/ [21 September 2007]; Doornbosch, Richard and Ronald Steenblik (September 2007). Biofuels: Is the Cure Worse than the Disease? OECD Roundtable Paper. Paris.

384. One consistent problem is that many environmental groups and many well-intentioned scientists presume idealistically that every step in a complex process will go perfectly and every involved party will behave properly based on social needs rather than private incentives. In the context of biofuels, for example, the utopian analysis assumed that producers would only use biological waste material and cellulosic ethanol made from twigs and dead leaves, rather than employing readily accessible corn and soy crops that would compete with food production. And the idealists assumed that farmers would continue to produce food crops as before, while growing suitable biofuel materials on marginal lands. However, as long as the prices of fossil fuels remain high, the prices of biofuels are likely to exceed the price of food crops per pound, and many farmers would prefer to get the higher prices for their work even if that choice may create national and international food shortages. The moral is that complexity, uncertainty, short-sightedness, and private interests or greed will always affect and

degrade efforts to produce desirable environmental conditions and public goods. Only a stargazing dreamer would support a program with as many obvious weaknesses and externalities as biofuels proliferation without carefully examining the negative aspects as well as the hypothetical good ones. See Streitfeld, David (9 April 2008). As Prices Rise, Farmers Spurn Conservation. *New York Times*, Online Edition; Rosenthal, Elisabeth (8 February 2008). Studies Deem Biofuels a Greenhouse Threat. *New York Times*, Online Edition.

385. Brewster, Rachel (2010). Stepping Stone or Stumbling Block: Incrementalism and National Climate Change Legislation. *Yale Law & Policy Review*, 28, 245–312.
386. *Id.*, p. 245.
387. Duhigg, Charles (23 November 2009). Sewers at Capacity, Waste Poisons Waterways. *New York Times*, Online Edition; Clean Water: Still Elusive (22 October 2009). *New York Times*, Online Edition.
388. Comprehensive Environmental Response, Compensation, and Liability Act, 42 U.S.C. §§ 9601–9675 (2000); Burlington Northern & Santa Fe Railway Company v. United States, 129 S. Ct. 1870 (2009).
389. Barringer, Felicity (1 January 2009). Move to Increase Logging on Oregon Land. *New York Times*, Online Edition; Associated Press (15 December 2008). Report: Endangered Species Decisions Tainted. *Seattle Times*, Online Edition.
390. Another analysis arguing for the adoption of multiple overlapping institutions is found in Reibstein, Richard (2009). Using the Tools of Pollution Prevention to Reduce Greenhouse Gas Emissions. *Environmental Law Reporter News & Analysis*, 39, 10851–10861.
391. Schiermeier, Quirin (2 April 2008). Climate Challenge Underestimated? Technology Will Not Automatically Come to Our Aid, Experts Warn. *Nature*, Online Edition; Pielke, Roger, Jr., Tom Wigley and Christopher Green (3 April 2008). Dangerous Assumptions. *Nature*, 452, 531–532; Pacala, Stephen and Robert Socolow (13 August 2004). Stabilization Wedges: Solving the Climate Problem for the Next 50 Years with Current Technologies. *Science*, 305(5686), 968–972.
392. Prager, Stewart C. (10 July 2011). How Seawater Can Power the World. *New York Times*, Online Edition.
393. *Id.*

394. Schelling, Thomas C. (2007). Climate Change: The Uncertainties, the Certainties, and What They Imply About Action. *Economists' Voice*, 4(3). Berkeley Electronic Press website, http://www.bepress.com/ev/vol4/iss3/art3/ [1 November 2007]; Stavins, Robert N. and Scott Barrett (2002). Increasing Participation and Compliance in International Climate Change Agreements. Harvard Kennedy School of Government Working Paper No. RWP02-031, pp. 18–20, 26.
395. Bosi, Martina *et al.* (May 2010). 10 Years of Experience in Carbon Finance: Insights from Working with the Kyoto Mechanisms. Report of the World Bank Carbon Finance Unit. World Bank Carbon Finance Unit website, http://siteresources.worldbank.org/INTCARBONFINANCE/Resources/10_Years_of_Experience_in_CF_August_2010.pdf/ [28 May 2010].
396. World Bank (2008). Carbon Finance at the World Bank: List of Funds. World Bank Carbon Finance Unit website, http://wbcarbonfinance.org/Router.cfm?Page=Funds&ItemID=3/ [10 August 2008].
397. Reuters (9 April 2010). World Bank Approves Loan for Coal-Fired Power Plant in South Africa. *Washington Post*, Online Edition.
398. Stewart, Richard B., Benedict Kingsbury and Bryce Rudik (eds.) (December 2009). Climate Finance for Limiting Emissions and Promoting Green Development: Mechanisms, Regulation and Governance. New York University School of Law Public Law Research Paper No. 09-66.
399. Nordhaus, William D. (September 2010). Carbon Taxes to Move Toward Fiscal Sustainability. *Economists' Voice*, 7(3). Berkeley Electronic Press website, http://www.bepress.com/ev/vol7/iss3/art3/.
400. *Id.*, p. 2.
401. *Id.*
402. *Id.*
403. Reich, Robert B. (2010). *Aftershock: The Next Economy and America's Future*, pp. 130–131. New York: Alfred A. Knopf.
404. Hansen, James (2009). *Storms of My Grandchildren*, pp. 209–222.
405. *Id.*, p. 172; —— (2007). How Can We Avert Dangerous Climate Change? Congressional Testimony. Previously available at http://pubs.giss.nasa.gov/authors/jhansen.html/ [27 September 2007]; —— (13 July 2006). The Threat to the Planet. *New York Review of Books*, 53(12), 13.

406. Rudolph, John Collins (23 August 2010). On Our Radar: Coal Plant Construction. *New York Times*, Online Edition.
407. For some of my previous writing on the intersections of law and cognitive psychology, see Latin, Howard (1994). Good Warnings, Bad Products, and Cognitive Limitations. *UCLA Law Review*, 41, 1193–1295. Reprinted in A. Bernstein (ed.) (1995). *A Products Liability Anthology*; —— (1994). Behavioral Criticisms of the Restatement (Third) of Torts: Products Liability. *Journal of Products and Toxics Liability*, 16, 209–220; —— (1985). Problem-Solving Behavior and Theories of Tort Liability. *California Law Review*, 73, 677–746.
408. The cap-and-trade advocates contend that their approach will achieve the same continuing incentives to reduce emissions because firms with low pollution control costs will be able to make a profit by selling their excess shares or allowances to firms with higher control costs. I am skeptical about this claim in practice because it is based on the dubious premise that most firms try to maximize their profits at all times, while I believe that many firms put a higher priority on ensuring income consistency and steady growth rather than profit maximization. The cap-and-trade system would require periodic auctions or distributions of the pollution shares, and many firms will hold onto their shares to increase their fiscal safety under conditions of uncertainty arising from evolving knowledge about global warming risks. In other words, the firms that can afford to sell the allowances may hold onto them for security and predictability reasons, or possibly to impede market entry by other firms, rather than following the idealized profit-maximization behavior that the cap-and-trade proponents assume will occur.
409. For a comprehensive assessment of this change in regulatory philosophies, see Latin, Howard (1985). Ideal versus Real Regulatory Efficiency: Implementation of Uniform Standards and 'Fine-Tuning' Regulatory Reforms. *Stanford Law Review*, 37, 1267–1332. Reprinted in R. Fischman, M. Lipeles and M. Squillace (eds.) (1996). *An Environmental Law Anthology*. Also reprinted in *Land Use & Environment Law Review* (1987).
410. *Id.*, pp. 1308–1309.
411. *Id.*, pp. 1304–1331; Wagner, Wendy E. (2000). The Triumph of Technology-Based Standards. *University of Illinois Law Review*, p. 83.

412. US EPA (15 December 2009). Endangerment and Cause or Contribute Findings for Greenhouse Gases Under Section 202(a) of the Clean Air Act. 40 CFR Chapter 1. *Federal Register*, 74(239), 66496–66546.
413. US EPA (23 December 2010). EPA Proposes Schedule to Address Greenhouse Gas Emissions from Refineries and Power Plants. US EPA website, http://www.epa.gov/climatechange/initiatives/index.html/ [31 January 2011].
414. US EPA (23 December 2010). Air Quality Planning and Standards: Addressing Greenhouse Gas Emissions. US EPA website, http://www.epa.gov/airquality/ghgsettlement.html/ [31 January 2011].
415. US EPA (3 June 2010). Prevention of Significant Deterioration and Title V Greenhouse Gas Tailoring Rule. *Federal Register*, 75(106), 31514.
416. *Id.*
417. *Id.*
418. *Id.*
419. *Id.*
420. "New sources as well as existing sources not already subject to title V that emit, or have the potential to emit, at least 100,000 tpy CO_2e will become subject to the PSD and title V requirements." (*Id.*) In addition, pollution "sources that emit or have the potential to emit at least 100,000 tpy CO_2e and that undertake a modification that increases net emissions of GHGs by at least 75,000 tpy CO_2e will also be subject to PSD requirements." (*Id.*)
421. *Id.*
422. *Id.*
423. Troutman Sanders LLP (July 2011). Washington Energy Report: House Committees Pass Bills to Slash EPA Funding and Delay Recent Clean Air Act Rules. Troutman Sanders LLP website, http://www.troutmansandersenergyreport.com/2011/07/house-committees-pass-bills-to-slash-epa-funding-and-delay-recent-clean-air-act-rules/ [26 July 2011]: Copy on file with the author.
424. Broder, John M. (13 June 2011). E.P.A. Plans Delay of Rule on Emissions. *New York Times*, Online Edition.
425. *Id.*
426. *Id.*
427. Broder, John M. (5 April 2011). White House Promises Veto of Anti-EPA Bill. *New York Times*, Online Edition; Broder, John M. (10 March 2011).

House Panel Votes to Strip EPA of Power to Regulate Greenhouse Gases. *New York Times*, Online Edition.

428. US EPA (25 October 2010). Regulatory Initiatives: EPA and NHTSA Propose Medium and Heavy Duty Vehicle Regulations to Reduce GHGs and Improve Fuel Efficiency. US EPA website, http://epa.gov/otaq/climate/regulations.htm/ [31 January 2011].
429. Hodas, David R. (2008). Ecosystem Subsidies of Fossil Fuels. Widener Law School Research Paper Series No. 08-37; Friedman, Thomas L. (30 April 2008). Dumb as We Wanna Be. *New York Times*, Online Edition; Mann, Roberta (2007). Another Day Older and Deeper in Debt: How Tax Incentives Encourage Burning Coal and the Consequences for Global Warming. *Global Business & Development Law Journal*, 20, 111–141; Andrews, Edmund L. (29 May 2007). Lawmakers Push for Big Subsidies for Coal Process. *New York Times*, Online Edition; Bradsher, Keith and David Barboza (11 June 2006). Pollution from Chinese Coal Casts a Global Shadow. *New York Times*, Online Edition; Gosseries, Axel (2004). Historical Emissions and Free-Riding. *Ethical Perspectives*, 11(1), 36; Koplow, Doug and John Dernbach (2001). Federal Fossil Fuel Subsidies and Greenhouse Gas Emissions: A Case Study of Increasing Transparency for Fiscal Policy. *Annual Review of Energy and the Environment*, 26, 361–381.
430. Schelling, Thomas C. (2007). Climate Change: The Uncertainties, the Certainties, and What They Imply About Action. *Economists' Voice*, 4(3). Berkeley Electronic Press website, http://www.bepress.com/ev/vol4/iss3/art3/ [1 November 2007]; Sandmo, Agnar (July 2006). Global Public Economics: Public Goods and Externalities. Norwegian School of Economics and Business Administration Working Paper; Gosseries, Axel (2004). Historical Emissions and Free-Riding. *Ethical Perspectives*, 11(1), 36.
431. US Environmental Protection Agency (31 March 2008). What Is the Toxics Release Inventory (TRI) Program? US EPA website.
432. MacDonald, Karen E. (2008). The European Pollutant Release and Transfer Register: A Case Study of Bosnia-Herzegovina. *European Journal of Law Reform*, 10(1), 21–49.
433. Vandenbergh, Michael P. and Anne C. Steinemann (2007). The Carbon-Neutral Individual. *New York University Law Review*, 82, 1673; Lin, Albert C.

(21 July 2008). Evangelizing Climate Change. Draft Working Paper. Social Science Research Network website, http://ssrn.com/abstract= 1142919/ [20 August 2008].

434. Farber, Daniel A. (2005). Controlling Pollution by Individuals and Other Dispersed Sources. *Environmental Law Reporter*, 35(10), 745.
435. US Environmental Protection Agency (2010). Greenhouse Gas Reporting Program. 40 CFR Part 98.
436. *Id.*
437. Berger, Matthew (29 January 2010). SEC Decision Requiring Disclosure of Climate Risks Could Have Broad Impact. Solve Climate: Daily Climate News and Analysis website, http://solveclimatenews.com/news/20100129/sec-decision-requiring-disclosure-climate-risks-could-have-broad-impact?page=2/ [7 September 2010].
438. Wheeler, David (November 2007). Moving Toward Consensus on Climate Policy: The Essential Role of Global Public Disclosure. Center for Global Development Working Paper No. 132.
439. Goulder, Lawrence H. and Ian W.H. Parry (April 2008). Instrument Choice in Environmental Policy. Resources for the Future Discussion Paper No. 08-07.
440. Associated Press (19 December 2010). 2010's World Gone Wild: Quakes, Floods, Blizzards. *New York Times*, Online Edition.
441. National Research Council (2010). *Assessment of Intraseasonal to Interannual Climate Prediction and Predictability*. Washington, D.C.: The National Academies Press; US EPA (2010). Climate Change Indicators in the United States. US EPA website, http://www.epa.gov/climatechange/indicators.html/ [29 April 2010]; Broder, John M. (19 May 2010). U.S. Science Body Urges Action on Climate. *New York Times*, Online Edition; UNSW Climate Change Research Centre (November 2009). The Copenhagen Diagnosis: Updating the World on the Latest Climate Science. http://www.copenhagendiagnosis.org/ [26 June 2011]; Pachauri, R.K. and A. Reisinger (eds.) (2008). *Intergovernmental Panel on Climate Change Fourth Assessment Report. Climate Change 2007: Synthesis Report*. Geneva: UN Intergovernmental Panel on Climate Change.

442. Sulzberger, A.G. (13 June 2011). Missouri: 2 Levees Are Breached. *New York Times*, Online Edition; Associated Press (6 June 2011). Hundreds Flee Flooding in Iowa Town. *New York Times*, Online Edition; Zabarenko, Deborah (19 May 2011). Floods, Droughts Are 'New Normal' of Extreme U.S. Weather Fueled by Climate Change, Scientists Say. *Huffington Post* website, http://www.huffingtonpost.com/2011/05/19/floods-droughts-extreme-weather-us_n_864046.html/ [26 June 2011]; Hauser, Christine (17 May 2011). Flooding Takes Economic Toll, and It's Hardly Done. *New York Times*, Online Edition.
443. Associated Press (22 June 2011). Deadly Storms over Midwest, South Subside. MSNBC website, http://www.msnbc.msn.com/id/14997870/ns/news/t/deadly-storms-over-midwest-south-subside/ [26 June 2011]; Associated Press (11 June 2010). 16 Dead, Dozens Missing in Ark. Floods. msnbc.com, file available from the author; Harless, William and Liz Robbins (4 May 2010). River Begins to Recede in Tennessee. *New York Times*, Online Edition; Dewan, Shaila (23 November 2009). In Mississippi Delta, a Promising Summer Washed Away by the Fall. *New York Times*, Online Edition.
444. Lacey, Marc (9 June 2011). More Evacuations in Arizona as Wildfire Surges. *New York Times*, Online Edition; msnbc.com staff (8 June 2011). Heat Wave Melts Records Across East Coast. MSNBC website, http://www.msnbc.msn.com/id/43315002/ns/weather/t/heat-wave-melts-records-across-east-coast/ [26 June 2011]; Associated Press (19 December 2010). 2010's World Gone Wild: Quakes, Floods, Blizzards. *New York Times*, Online Edition.
445. Goodnough, Abby (2 June 2011). Massachusetts Begins Cleanup After Tornadoes. *New York Times*, Online Edition; O'Connor, Anahad (25 May 2011). At Least 14 People Are Killed in Storms in 3 States. *New York Times*, Online Edition; Stelter, Brian and A.G. Sulzberger (24 May 2011). Searches Restart in Deadliest U.S. Tornado in 60 Years. *New York Times*, Online Edition.
446. Paulson, Stephen K. (22 June 2011). Heavy Snows Spoil Weekend Holiday Plans in West. MSNBC website, http://www.msnbc.msn.com/id/43192780/ns/travel-news/t/heavy-snows-spoil-weekend-holiday-plans-west/ [26 June 2011]; Johnson, Kirk and Jesse McKinley (21 May

2011). Record Snowpacks Could Threaten Western States. *New York Times*, Online Edition.

447. Galbraith, Kate (18 June 2011). Amid Texas Drought, High-Stakes Battle over Water. *New York Times*, Online Edition; Krugman, Paul (6 February 2011). Droughts, Floods and Food. *New York Times*, Online Edition; Barringer, Felicity (27 September 2010). Water Use in Southwest Heads for a Day of Reckoning. *New York Times*, Online Edition.
448. Robertson, Campbell (1 June 2011). As Hurricane Season Begins, Heavy Storms Are Expected. *New York Times*, Online Edition; Hauser, Christine (1 June 2011). In Wake of Natural Disasters, Insurers Brace for Big Losses. *New York Times*, Online Edition.
449. Metz, Bert (2010). *Controlling Climate Change*. Cambridge: Cambridge University Press; National Research Council (2010). *Assessment of Intraseasonal to Interannual Climate Prediction and Predictability*. Washington, D.C.: The National Academies Press; Rummukainen, Markku *et al.* (May 2010). Physical Climate Science Since IPCC AR4: A Brief Update on New Findings Between 2007 and April 2010. Nordic Council of Ministers.
450. Hauser, Christine (1 June 2011). In Wake of Natural Disasters, Insurers Brace for Big Losses. *New York Times*, Online Edition; International Bank for Reconstruction and Development/World Bank and United Nations (November 2010). Natural Hazards, UnNatural Disasters: The Economics of Effective Prevention. World Bank website, http://siteresources.worldbank.org/INFOSHOP1/Resources/naturalhazards.pdf/ [26 June 2011].
451. Krugman, Paul (23 May 2010). The Old Enemies. *New York Times*, Online Edition.
452. International Centre for Trade and Sustainable Development (21 February 2011). Trillions in Global Investments at Risk Due to Climate Change: Study. ICTSD website, http://ictsd.org/i/news/biores/101276/ [27 June 2011].
453. Lewandowsky, Stephan (28 April 2011). Acceptance of Science and Ideology. Shaping Tomorrow's World website, http://www.shapingtomorrowsworld.org/ideologyScience.html/ [27 June 2011]; Revkin, Andrew C. (28 April 2011). On Birth Certificates, Climate Risk and an Inconvenient Mind. *New York Times*, Online Edition; Hamilton, Clive (28 October 2010). Why We Resist the Truth About Climate

Change. http://www.clivehamilton.net.au/cms/media/why_we_resist_the_truth_about_climate_change.pdf/ [27 June 2011]; Biber, Eric (2009). Climate Change and Backlash. *New York University Environmental Law Journal*, 17, 1295–1366.

454. Lewandowsky, Stephan (28 April 2011). The Truth Is Out There. The Drum Opinion. ABC website, http://www.abc.net.au/unleashed/266186.html/ [27 June 2011]; Revkin, Andrew C. (28 April 2011). On Birth Certificates, Climate Risk and an Inconvenient Mind. *New York Times*, Online Edition.

455. Norgaard, Kari Marie (2011). *Living in Denial: Climate Change, Emotions, and Everyday Life*. Cambridge, MA: MIT Press; Revkin, Andrew C. (8 May 2011). Is a Human "Here and Now" Bias Clouding Climate Reasoning? *New York Times*, Online Edition.

456. Kaufman, Leslie (15 April 2011). G.O.P. Push in States to Deregulate Environment. *New York Times*, Online Edition.

457. Hansen, James (2009). *Storms of My Grandchildren*, pp. 168, 185–186, 220.

458. Kaufman, Leslie (9 June 2011). Americans Still Split on Global Warming, Poll Shows. *New York Times*, Online Edition; Hamilton, Clive (28 October 2010). Why We Resist the Truth About Climate Change. http://www.clivehamilton.net.au/cms/media/why_we_resist_the_truth_about_climate_change.pdf/ [27 June 2011]; Harshaw, Tobin (23 October 2009). Are Americans Cooling on Global Warming? *New York Times*, Online Edition.

Index